International Mathematical Congresses

AN ILLUSTRATED HISTORY
1893–1986

REVISED EDITION

Including ICM 1986

Donald J. Albers
G.L. Alexanderson
Constance Reid

Springer-Verlag

D. Albers
Math Department
Menlo College
Menlo Park, CA 94025
U.S.A.

G. Alexanderson
Department of Math
University of Santa Clara
Santa Clara, CA 95053
U.S.A.

Constance Reid
70 Piedmont Street
San Francisco, CA 94117
U.S.A.

Cover design by Kathy Avanzino Barone

AMS Classification: 01A60, 01A65, 01A70, 01A99

Library of Congress Cataloging in Publication Data
Albers, Donald J., 1941–
    International mathematical congresses.
    1. International Congress of Mathematicians—
History.   2. Mathematics—Congresses—History.
I. Alexanderson, Gerald L.   II. Reid, Constance.
III. International Congress of Mathematicians.
IV. Title.
QA1.A464   1987   510′.6′01   86-31401

Typesetting in Times Roman by Acme Type Co., Inc., San Mateo, CA
Printed and bound by Quinn-Woodbine, Woodbine, New Jersey.
Printed in the United States of America.

9 8 7 6 5 4 3 2 1

ISBN 0-387-96479-7 Springer-Verlag New York Berlin Heidelberg
ISBN 3-540-96479-7 Springer-Verlag Berlin Heidelberg New York

Table of Contents

International Mathematical Congresses

AN ILLUSTRATED HISTORY
1893–1986

Chicago 1893

Felix Klein

As part of the Chicago World's Columbian Exposition, celebrating the 400th anniversary of Columbus's discovery of America, a series of scientific and philosophical congresses was scheduled, among them a "world congress" of mathematicians and astronomers. Felix Klein of Göttingen brought a number of contributions from European mathematicians and opened the Congress with a brief address on "The Present State of Mathematics," in which he emphasized that "what was formerly begun by a single mastermind, we now must seek to accomplish by united efforts and cooperation."

After Klein's address mathematicians and astronomers separated into their own sections. Forty-five mathematicians attended the mathematics section, three of them, in addition to Klein, coming from abroad. The American participants were a far from homogeneous group in their academic backgrounds: 23 listed Ph.D.'s; 9, M.A.'s; 4, A.B.'s; and 6, no degrees. Two M.D.'s also attended.

At the final session W. E. Story of Clark University, the President of the Congress, expressed for the group the specific indebtedness of the mathematics section to Professor Klein and the general indebtedness of American mathematics "to the influence and inspiration of German universities and mathematicians."

Klein had agreed to hold a colloquium at Northwestern University after the close of the Congress for those who wished to attend. There he delivered, "freely in the English language," a series of lectures on recent developments in mathematics with which he had been closely connected as a result of his own work or of personal observation.

The "Evanston Colloquium" was the first mathematical colloquium on American soil. Among the 25 mathematicians who attended, there was one woman, Mary F. Winston, A.B., Honorary Fellow, University of Chicago. Klein suggested that she come to Göttingen for further study. A gift of $500 from Christine Ladd Franklin, the first woman to earn a Ph.D. in mathematics in America (awarded 44 years later), made the trip possible.

Art Palace, site of the World Congress of Mathematicians and now the home of the Museum of Science and Industry.

> But our mathematicians must go further still. They must form international unions, and I trust that this . . . World Congress will be a step in that direction.
>
> — Felix Klein, 1893

The Present State of Mathematics

. . . When we contemplate the development of mathematics in this nineteenth century, we find something similar to what has taken place in other sciences. The famous investigators of the preceding period . . . were each great enough to embrace all branches of mathematics and its applications. In particular, astronomy and mathematics were in their time regarded as inseparable.

With the succeeding generation, however, the tendency to specialisation manifests itself. . . . But the developing science departs at the same time more and more from its original scope and purpose and threatens to sacrifice its earlier unity and to split into diverse branches. In the same proportion the attention bestowed upon it by the general scientific public diminishes. It became almost the custom to regard modern mathematical speculation as something having no general interest or importance. . . .

This is a picture of the past. I wish on the present occasion to state and to emphasize that in the last two decades a marked improvement from within has asserted itself in our science, with constantly increasing success.

The matter has been found simpler than was at first believed. It appears indeed that the different branches of mathematics have actually developed not in opposite, but in parallel directions, that it is possible to combine their results into general conceptions. . . .

Speaking, as I do, under the influence of our Göttingen tradition, and dominated somewhat, perhaps, by the great name of Gauss, I may be pardoned if I characterise the tendency that has been outlined in these remarks as a return to the general Gaussian programme.

— FELIX KLEIN

Contributors to the Congress

Papers reviewing recent developments in various branches of mathematics were contributed by the following mathematicians, although only those whose names are starred were personally present at the Congress:

Oskar Bolza, Chicago*; Heinrich Burkhardt, Göttingen; Alfredo Capelli, Naples; Frank N. Cole, Michigan; Walther v. Dyck, Munich; William H. Echols, Virginia; Henry T. Eddy, Indiana*; Robert Fricke, Göttingen; G. B. Halsted, Texas*; Lothar Heffter, Giessen; Charles Hermite, Paris; David Hilbert, Königsberg.

Adolf Hurwitz, Zürich; Felix Klein, Göttingen*; Martin Krause, Dresden; Emile Lemoine, Paris; Matyas Lerch, Prague; Alexander Macfarlane, Texas*; Artemas Martin, Washington D.C*; Heinrich Maschke, Chicago*; Franz Meyer, Clausthal; Hermann Minkowski, Bonn; E. H. Moore, Chicago*; Eugen Netto, Giessen; Max Noether, Erlangen; Maurice d'Ocagne, Paris; Bernard Paladini, Pisa*; Joseph de Perott, Massachusetts.

T. M. Pervouchine, Kazan; Salvatore Pincherle, Bologna; Alfred Pringsheim, Munich; Albert M. Sawin, Wisconsin; Victor Schlegel, Hagen; Arthur Schoenflies, Göttingen; Irving Stringham, California; Eduard Study, Marburg*; Henry Taber, Massachusetts*; Heinrich Weber, Göttingen; Edouard Weyr, Prague.

The University of Chicago's **E. H. Moore** with his colleagues, **Oskar Bolza** and **Heinrich Maschke**, both former students of Klein.

Zurich 1897

C. F. Geiser

O n November 17, 1896, Hermann Minkowski, a professor at the Zürich Polytechnikum, or Eidgenössische Technische Hochschule, better known today as the ETH, added a postscript to his letter to David Hilbert: "It just occurs to me — for several days we have been meeting in regard to the international mathematical congress."

An announcement of the proposed congress went out at the beginning of the new year above the names of C. F. Geiser, ETH, the nephew of Jacob Steiner, and 21 distinguished mathematicians, 10 of whom were from Zürich:

"After considerable correspondence . . ., the question of the place of holding the Congress has been decided in favor of Switzerland, as a country peculiarly adapted by situation, relations, and traditions of promoting international interests."

By January 27 Minkowski was writing again to Hilbert that the arrangements for excursions and other congress entertainments had been made: "Here, naturally, once more Science comes last."

On Sunday, August 8, 1897, the members of the Welcoming Committee, headed by Adolf Hurwitz, ETH, were on hand in the Zürich *Bahnhof* to greet arriving mathematicians and transport them to their lodgings.

"It is true that the great conceptions of our science generally develop and mature in the silent study of the scholar," Hurwitz conceded at the welcoming banquet that evening. "No science, unless perhaps philosophy, has such a brooding and solitary character as mathematics. But nevertheless there lives in the breast of the mathematician the necessity for communication, for conversation with colleagues."

Such stimulating scientific intercourse was to be the primary goal of the Congress, which — although following the World Congress in Chicago — was to be accepted as the First International Congress of Mathematicians.

The next morning the three-day meeting was opened by Geiser, who had been elected president by acclamation. Ferdinand Rudio, the General Secretary, addressed the session on the functions and the organization of international mathematical congresses. Those present voted that his talk should appear in the published proceedings in French as well as German.

Ferdinand Rudio

Plenary lectures at the first official international congress of mathematicians were to be delivered by Henri Poincaré, Paris; Adolf Hurwitz, Zürich; Giuseppe Peano, Turin; and Felix Klein, Göttingen — two on the first and two on the last day of the Congress. In Poincaré's absence, because of illness, Jérôme Franel, the ETH's "Professor für Mathematik in französische Sprache," read his paper. The middle day of the Congress was devoted to five section meetings. On the final day members and their guests went by train to the summit of the Uetli for a farewell banquet. It was reported that many "remained till a late hour in the evening, enjoying the moonlight landscape that lay before and beneath them."

Der Kongress hat den Zweck:
Die persönlichen Beziehungen zwischen den Mathematikern
der verschiedenen Länder zur Fördern. . .
—Art. 1, Die Reglement

William Fogg Osgood Reports:

Professor Rudio pointed out some of the more important ends that international congresses may serve.

"Viribus unitis! Let this be our watchword! With the union of our forces it will be possible to accomplish tasks that hitherto, in the absence of cooperation, could not even be attempted. To give an illustration. Here on Swiss soil you will pardon me if I take as such an edition of Euler's works, a debt of honor, which the mathematical world has hitherto been unable to discharge. An important piece of preliminary work which must precede the direct attack of this great task has now been performed, I mean the work of our American colleague, Mr. Hagen, who, as you know, published a year ago a complete list of Euler's works. . . . an edition of these works is now no longer a chimerical project, nay, requires perhaps nothing more than international moral support."

Professor Rudio then mentioned without comment several further suggestions that the Committee has received: a directory, to appear if possible annually, of all the mathematicians of the world, with a statement of their special fields of work; a bibliographical and literary dictionary of all mathematicians now living, together with their portraits. . . . The speaker closed by emphasizing strongly the need of cooperation in bibliographical

Adolf Hurwitz

matters. . . . He suggested a publication that should contain the exact title of every paper that appears in the world and that should be issued in such a form that the information it contained would be sent out within a month, perhaps even within a week of the receipt of the same by the editors. Such a publication, he pointed out, would have to be preceded by a classification of mathematical literature which should have been universally adopted.

POINCARÉ: The First Lecture

"Sur les rapports de l'analyse pure et de la physique mathématique"

. . . The combinations that can be formed with numbers and symbols are an infinite multitude. In this thicket how shall we choose those that are worthy of our attention? Shall we be guided only by whimsy? . . .[This] would undoubtedly carry us far from each other, and we would rapidly cease to understand each other. But that is only the minor side of the problem. Not only will physics perhaps prevent us from getting lost, but it will also protect us from a more fearsome danger . . .turning around forever in circles. History [shows that] physics has not only forced us to choose [from the multitude of problems which arise], but it has also imposed on us directions that would never have been dreamed of otherwise. . . . What could be more useful!

KLEIN: The Final Lecture

"Zür Frage des höheren mathematischen Unterrichtes"
. . . Nor shall I leave untouched the general question of the close relationship between mathematics and applications, a question which M. Poincaré made the subject of his brilliant exposé. No one is more convinced than I of the importance of this close relationship; however, before the assembled mathematical congress, it seems appropriate to me to emphasize another aspect, i.e., that there is such a thing as pure mathematics — that it is, in fact, the core of our science, the prosperity of which is the prerequisite of all other mathematical activities if they are not quickly to decline to a lower level. Therefore, let me speak of the training of those few who are destined to carry pure mathematics into the future, the true mathematical scholars.

VIRIBUS UNITIS! SEI UNSERE LOSUNG

Paris 1900

Charles Hermite

International congresses are now traditionally held every four years; however, the constitution set down at Zürich provided three to five years between congresses, since it had already been decided that the Second International Congress of Mathematicians would be held in Paris in 1900 to coincide with that city's *Exposition universelle* and the end of the nineteenth century. A thousand mathematicians had indicated that they would attend; but for a multiplicity of reasons, including fear of crowds and high prices, only 229 actually put in an appearance in August at Paris, just 21 more than had been at Zürich.

Under the presidency of Henri Poincaré, the Congress opened on the grounds of the *Exposition* and then adjourned to the more scholarly environment of the Sorbonne. As at Zürich there were four principal speakers: Moritz Cantor, Heidelberg; Gösta Mittag-Leffler, Stockholm; Vito Volterra, Rome; and Poincaré himself. It was, however, a talk by David Hilbert entitled simply "Mathematische Probleme," presented to a combined meeting of the sections on History and Pedagogy, that was to give the Paris congress its place in the history of mathematics.

The congress, which lasted for seven days as compared to the three at Zürich, was beset by non-mathematical problems. In fact—

"The arrangements excited a good deal of criticism," reported Charlotte Angas Scott, Bryn Mawr, in *The AMS Bulletin*. "... There is no doubt that a smaller town lends itself best to such a gathering; it is not so much that there is less division of interests, as that the members are more in evidence, and so have a better chance of *realizing* one another."

There was also the problem of language. While French and German had both played a role at Zürich, the emphasis at Paris was exclusively on French. All of the plenary addresses and most of the sectional papers were in that language, "possibly out of compliment to our hosts," hazarded Scott.

Later, reviewing the published *Comptes Rendus* of the Congress, she was surprised to find that all the papers appeared in French with the exception of one in Italian and eight in English: "But there were certainly some in German." One paper delivered in German was, of course, that of Hilbert. In the *Comptes Rendus* it was placed among the plenary talks "en raison de sa grande importance"—but in French translation and entitled "Problèmes futurs des mathématiques."

Papers were delivered in English by mathematicians of several nationalities, including Japanese. Alexander

Henri Poincaré

Vassilief, Kazan, pointed out that, judging by the ardor with which the latter were pursuing the subject, there would soon be a sizable body of mathematical literature in Japanese. He proposed a resolution, which was adopted, that scholarly academies and societies study means of remedying the variety of languages employed in scientific literature. Although Esperanto was not mentioned, it was the language he had in mind.

Mittag-Leffler, who since the Franco-Prussian war had endeavored to bring together French and German mathematicians by publishing them side by side in *Acta Mathematica,* began his address by describing how three years after the war he had come to Paris to follow the lectures of Hermite.

"I shall never forget my astonishment at the first words which he addressed to me: 'You have made a mistake, sir,' he said. 'You ought to follow the lectures of Weierstrass in Berlin. He is the master of us all.' Hermite was a Frenchman and a patriot. I realized at the same time to what degree he was a mathematician."

Moritz Cantor

Lieber Freund! Hilbert Asks for Advice

He would like to make a speech which would be appropriate to the significance of the occasion. In his New Year's letter to Minkowski, he mentioned receiving the invitation and recalled the two speeches from the first International Congress which had so impressed him — the scintillating but technical lecture by Hurwitz on the history of the modern theory of functions and the popular discourse by Poincaré on the reciprocal relationship existing between analysis and physics. He had always wanted to reply to Poincaré with a defense of mathematics for its own sake, but he also had another idea. He had frequently reflected upon the importance of individual problems in the development of mathematics. Perhaps he could discuss the direction of mathematics in the coming century in terms of certain important problems on which mathematicians should concentrate their efforts. What was Minkowski's opinion? . . .

On January 5, 1900, Minkowski wrote. . . .

"I have re-read Poincaré's lecture . . . and I find that all his statements are expressed in such a mild form that one cannot take exception to them. . . . Since you will be speaking before specialists, I find a lecture like the one by Hurwitz better than a mere chat like that of Poincaré. . . .

"Most alluring would be the attempt at a look into the future and a listing of the problems on which mathematicians should try themselves during the coming century. With such a subject you could have people talking about your lecture decades later."

There was no reply from Hilbert. . . .

On March 29 he consulted Hurwitz.

". . . I am hesitating about a subject. . . . The best would be a view into the future. What do you think about the likely direction in which mathematics will develop during the next century? It would be very interesting and instructive to hear your opinion about that."

From HILBERT by Constance Reid (Springer 1970)

Hermann Minkowski

Hilbert on proofs of the impossibility of a solution of a mathematical problem

David Hilbert c. 1900

"Who of us would not be glad to lift the veil behind which the future lies hidden; to cast a glance at the next advances of our science and at the secrets of its development during future centuries?" Thus David Hilbert began his historic speech on mathematical problems at the International Congress of Mathematicians in 1900.

It seemed to him that "if we would obtain an idea of the probable development of mathematical knowledge in the future, we must let the unsettled questions pass before our minds and look over the problems which the science of today sets. . . ." In the published version of his talk he listed, "tentatively as it were," the 23 problems which appear on this page, "from the discussion of which an advancement of science may be expected." In his actual talk however, he had time to treat only 10.*

Excerpted below is that portion of his talk in which he discusses the fact that a mathematical problem may on occasion be satisfactorily solved by a proof of the impossibility of a solution:

"Occasionally it happens that we seek the solution under insufficient hypotheses or in an incorrect sense, and for this reason do not succeed. The problem then

The Future Problems of Mathematics

1. Cantor's problem of the cardinal number of the continuum.
2. The compatibility of the arithmetical axioms.
3. The equality of the volumes of two tetrahedra of equal bases and equal altitudes.
4. The problem of the straight line as the shortest distance between two points.
5. Lie's concept of a continuous group of transformations without the assumption of the differentiability of the functions defining the group.
6. The mathematical treatment of the axioms of physics.
7. The irrationality and the transcendence of certain numbers.
8. Problems of prime numbers (including the Riemann hypothesis).
9. The proof of the most general law of reciprocity in any number field.
10. The determination of the solvability of a Diophantine equation.
11. The problem of quadratic forms with any algebraic numerical coefficients.
12. The extension of Kronecker's theorem of Abelian fields to any algebraic realm of rationality.
13. The proof of the impossibility of the solution of the general equation of the 7th degree by means of functions of only two arguments.
14. The proof of the finiteness of certain complete systems of functions.
15. A rigorous foundation of Schubert's enumerative calculus.
16. The problem of the topology of algebraic curves and surfaces.
17. The expression of definite forms by squares.
18. The building up of space from congruent polyhedra.
19. The determination of whether the solutions of "regular" problems in the calculus of variations are necessarily analytic.
20. The general problem of boundary values.
21. The proof of the existence of linear differential equations having a prescribed monodromic group.
22. Uniformization of analytic relations by means of automorphic functions.
23. The further development of the methods of the calculus of variations.

Problems 1, 2, 6, 7, 8, 13, 16, 19, 21, 22.

arises: to show the impossibility of the solution under the given hypotheses, or in the sense contemplated. Such proofs of impossibility were effected by the ancients, for instance when they showed that the ratio of the hypotenuse to the side of an isosceles right triangle is irrational. In later mathematics, the question as to the impossibility of certain solutions plays a preëminent part, and we perceive in this way that old and difficult problems, such as the proof of the axiom of parallels, the squaring of the circle, or the solution of equations of the fifth degree by radicals have finally found fully satisfactory and rigorous solutions, although in another sense than that originally intended. It is probably this important fact along with other philosophical reasons that gives rise to the conviction (which every mathematician shares, but which no one has yet supported by a proof) that every definite mathematical problem must necessarily be susceptible of an exact settlement, either in the form of an actual answer to the question asked, or by the proof of the impossibility of its solution and therewith the necessary failure of all attempts. Take any definite unsolved problem, such as the question as to the irrationality of the Euler-Mascheroni constant C, or the existence of an infinite number of prime numbers of the form $2^n + 1$. However

unapproachable these problems may seem to us and however helpless we stand before them, we have, nevertheless, the firm conviction that their solution must follow by a finite number of purely logical processes.

"Is this axiom of the solvability of every problem a peculiarity characteristic of mathematical thought alone, or is it possibly a general law inherent in the nature of the mind, that all questions which it asks must be answerable? For in other sciences also one meets old problems which have been settled in a manner most satisfactory and most useful to science by the proof of their impossibility. I instance the problem of perpetual motion. After seeking in vain for the construction of a perpetual motion machine, the relations were investigated which must subsist between the forces of nature if such a machine is to be possible; and this inverted question led to the discovery of the law of the conservation of energy, which, again, explained the impossibility of perpetual motion in the sense originally intended.

"This conviction of the solvability of every mathematical problem is a powerful incentive to the worker. We hear within us the perpetual call: There is the problem. Seek its solution. You can find it by pure reason, for in mathematics there is no *ignorabimus*.

. . . And the Ladies of the Club . . .

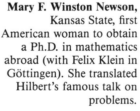

Mary F. Winston Newson, Kansas State, first American woman to obtain a Ph.D. in mathematics abroad (with Felix Klein in Göttingen). She translated Hilbert's famous talk on problems.

Charlotte A. Scott, Bryn Mawr, took her Ph.D. at the University of London in 1885 and became head of the Mathematics Department at Bryn Mawr. She reported Hilbert's talk in *The AMS Bulletin.*

Heidelberg 1904

Heinrich Weber

"As Jacobi's words sent out sparks and fired the enthusiasm of his audiences, so did the contents and style of his writing call forth a clear and persisting echo in the minds of the new generation of mathematicians," Königsberger told his audience. "In this sense we may also count among his students the two foremost representatives of mathematical scholarship in the second half of the last century—Hermite and Weierstrass. And all of us, students of these two outstanding scholars who have passed on Jacobi's words and views with awe and veneration for an oracle of mathematical mysteries—all of us who are here assembled to honor the memory of that great master are indeed the students of Jacobi."

The hall of the university museum, which was used for the opening session, was afterwards devoted to an innovative and extensive exhibition of mathematical literature, apparatus, and models of the recent decade. The exhibition was formally inaugurated with three general lectures, including a discussion and display of Leibniz's original calculating machine by Carl Runge, Hannover. He pointed out that while the machine probably never worked with entire accuracy because of faulty mechanical construction, the principles underlying its construction were the same as those in successful contemporary machines.

The invitation extended in Paris by the Deutsche Mathematiker-Vereinigung had specified Baden-Baden, a spa in the Black Forest, as the site of the Third International Congress; but within a year the members had decided to substitute the scenic university town of Heidelberg.

Heinrich Weber, Strassburg, organized the event, and he and his committee made a tremendous effort to increase attendance. In addition to general invitations extended to academies and scientific societies, postcards were sent to more than two thousand individuals whose names had been culled from the subscription lists of sixteen journals. The result was that while Paris had exceeded Zürich by only 21 mathematicians, Heidelberg drew 107 more than Paris. It appears, however, that not a single one of the participating members was female.

The organizing committee had early decided on one plenary address in English (A. G. Greenhill), one in French (P. Painlevé), one in German (W. Wirtinger, Vienna), and one in Italian (C. Segre). It had also decided *ganz besonders aber alle* that the lectures should be published in the language in which they were given.

Weber, as President, opened the Congress by reminding the members of the many losses mathematics had suffered in recent years, among them Weierstrass (1897) and Hermite (1901). The year of the Congress was the 100th anniversary of the birth of C. G. J. Jacobi, and a special observance was planned. Following Weber's address, Leo Königsberger, Heidelberg, delivered the featured *Gedächtnisrede*.

Leo Königsberger

In the absence of Walther v. Dyck, chairman of the commission for a mathematical encyclopedia, Felix Klein presented the first German volume to the Congress while Julius Molk, Nancy, presented the French volume which, he explained, although following closely the German version, preserved traditional French usage.

> . . . of all the most beautiful spots on earth which
> our German Fatherland calls its own, there is only
> one Heidelberg!
>
> — Heinrich Weber, 1904

Henri Fehr, founder of *L'Enseignement Mathématique,* circulated a questionnaire on the methods of work practiced by mathematicians, prepared by the editors in collaboration with two Geneva psychologists.

"The first group [of questions] deals with the psychological side," *The AMS Bulletin* reported to its readers, "embracing such topics as the genesis of a taste for the science, influence of heredity, relative interest in pure and applied mathematics, the roles of chance and inspiration, unconscious cerebration, practice of assimilating existing literature as preparation for an investigation or of leaving the mind unfettered, advice to beginners. The second group relates to the mathematician's mode of life, hours of work, regularity of habits, avocations, physical exercise, requisite vacations. The third group invites miscellaneous data of value for the inquiry, effect of narcotics, favorite attitude and appliances while at work, nature of mental images, personal recollections of the habits of deceased mathematicians, etc."

On their visit to the Grand Ducal Palace, although the Duke was not present to greet them, the mathematicians were met at the station by an assemblage of fire companies, veterans' organizations, musical societies and school children, and "had the distinction — unusual for mathematicians" of proceeding so accompanied to the

Paul Painlevé

castle gates. All in all, the Congress seems to have been a week-long extravaganza of talks, exhibitions, excursions, fireworks — even the Pythagorean Theorem blazing in the sky!

"One could have wished only for an increase of the day's normal twenty-four hours. . . ."

Rome 1908

Vito Volterra

The Fourth International Congress of Mathematicians convened in the spring of 1908 in the ancient center of the civilized world. In fact, at times, according to Henri Fehr, Geneva — by this time an inveterate attender of the congresses — it seemed that the attractions of Rome itself diverted the members of the Congress from the presentations being made at the Palazzo Corsini, not so centrally located for sightseeing as it might have been.

The Italians departed from several of the tenuously established traditions of international congresses: first in choosing spring for their congress and secondly in more than doubling the number of mathematicians invited to give plenary addresses. Among these Poincaré and Darboux were asked to represent France; Hilbert and Klein, Germany; Volterra and Veronese, Italy; Simon Newcomb, United States, the first American to deliver a plenary address. (He spoke in French.) Klein and Hilbert had to decline; and, once again, Poincaré was unable to present his lecture in person.

"It was at Rome . . .," C. G. Darwin was later to recall, "that the first dark shadow fell on us of that illness which has now terminated so fatally. You all remember the dismay . . . when the word passed from man to man 'Poincaré is ill.' "

The opening session of the Congress took place in the Capitol in the presence of the King of Italy and featured the address by Vito Volterra, which traced the development of mathematics in Italy from the *Risorgimento,* "in which the whole life of the nation was renewed and the universities reopened," to the beginning of the present century.

Another feature of the opening session was the awarding of the *Guccia Medaglia.* The prize — a medal in gold accompanied by 3000 francs — had been offered at Heidelberg by the *Circolo Matematico di Palermo* for "A memoir making an important advance in the theory of algebraic twisted curves."

"It is well known that [in recent years] the theory of algebraic twisted curves has been somewhat neglected," *The AMS Bulletin* had pointed out at the time.

Three memoirs, ranging from 16 to 234 pages, had been submitted; but none of the three was considered worthy by the prize committee (M. Noether, H. Poincaré, and C. Segre) — a possibility that had apparently been foreseen in the guidelines set down for the award. As an alternative, the *Guccia Medaglia* went by the unanimous vote of the committee to Francesco Severi, Padua, for his many investigations on the geometry of algebraic surfaces. It was the first time that a prize of any kind had been awarded at an international congress.

Max Noether brought his daughter, **Emmy,** as a guest.

> **But it is with sure faith that we face the future, hoping for the constant and harmonious development of Italian mathematical thought in union with that of other nations.**
>
> **—Vito Volterra, 1908**

Francesco Severi

After the *Inaugurazione del Congresso,* subsequent sessions under President Pietro Blaserna, Rome, the President of the *Accademia dei Lincei,* took place in the quarters of the Accademia at the Palazzo Corsini.

Almost as many of the papers were presented in French (51) as in Italian (53).

"Which shows," Henri Fehr concluded, "not only that the French school was brilliantly represented, but that a large number of scholars possess a knowledge of this language, which continues to maintain its place as the international language of the Congress."

David Eugene Smith

Simon Newcomb

D. E. Smith Presents Some "Advanced Views" on High School Math in the US:

Mathematics I. Required. Boys and girls together. Transition from special to general forms. A combination of algebra, concrete geometry and arithmetic.

Mathematics II. Required. Boys and girls in separate classes, with the possible proviso that girls may elect to enter the classes for boys. Elementary algebra combined as closely as possible with the first three books of plane geometry, each presupposing Mathematics I.

Mathematics III. Required for boys; elective for girls. Separate classes. Elementary algebra completed, plane geometry completed, mechanical drawing continued, and so on.

Mathematics IV. Elective for both boys and girls. Separate classes preferred. A course leading to mechanics and physiography and introducing the spherical triangle, the elements of analytic geometry and curve tracing generally.

Mathematics V (a). Elective for both boys and girls. Separate classes preferred. A half year of work in the calculus and its applications. It is possible that a parallel culture course for girls may be arranged, involving the elements of mathematical and descriptive astronomy.

Mathematics V (b). Required for boys, elective for girls. Separate classes. Commercial arithmetic. For the girls, particular attention to the arithmetic of the various branches of Domestic Economy.

Cambridge 1912

Arthur Cayley

The Fifth International Congress of Mathematicians was held in the English university city of Cambridge a few weeks after the sudden and premature death of Henri Poincaré. Even the English skies seemed to mourn the loss of the great mathematician, for it rained almost steadily throughout the Congress.

Attendance had increased in the fifteen years since Zürich, and a total of 708 guests descended upon the city, 574 described in the *Proceedings* as "effective members." Lodgings for the visitors were provided in the Cambridge colleges, men in the men's colleges and their accompanying wives and daughters, as well as women mathematicians, in the women's colleges. The "social report" on the Congress by Elizabeth B. Cowley, Vassar, which appeared in *The AMS Bulletin,* was, however, enthusiastic about "the opportunity—so freely and cordially offered to the members of the Congress—to live in these historical old buildings and to share, in a degree, the life of the University."

Although the English did not go so far as the Italians in expanding the roster of plenary speakers, they did double the number of speakers from English-speaking countries. Emile Borel represented French mathematics and Edmund Landau, German. An invitation to give a plenary address went to a representative of Russian science: Prince B. Galitzen of the St. Petersburg academy. Speakers were evenly divided between pure mathematics and applied. In addition, the Section on Applications was divided into three subsections that reflected British interests: Engineering Mathematics; Statistical, Economic,

Émile Borel

and Actuarial Mathematics; and Mathematical Astronomy.

Sir C. G. Darwin of the Cambridge Philosophical Society presided over the Congress. Both he and other speakers took the opportunity to emphasize to their guests that English mathematicians had finally broken out of their long isolation from the mathematicians of the continent.

"There was a time not long ago," conceded E. B. Elliott, Oxford, "when British mathematicians may have been thought too self-centered. If the judgment were ever correct, it is no longer. We are alive to what is being done elsewhere, and now aim at cooperation. . . ."

Progress was reported on several projects supported by past congresses, most especially the publication of the works of Euler, which had reached five volumes and was proceeding at an encouraging rate. The International Committee on the Teaching of Mathematics had accumulated some 150 volumes and 300 articles from all parts of the world: "None of us . . . in Rome could even imagine what an immense labor was to be undertaken when Dr. D. E. Smith proposed a comparative investigation of mathematics teaching," commented Walther v. Dyck, Munich, moving that Smith be added to the committee—at its request.

On the last day of the Congress, at the suggestion of Samuel Dickstein, Warsaw, a group of members made a pilgrimage to the Mill Road Cemetery, where Arthur Cayley is buried, for the purpose of laying a wreath on his grave. "This has touched the hearts of our University," Darwin later commented. Arrangements were made for a silver wreath to be purchased and displayed "where it will remain as a permanent memorial of the recognition accorded by the mathematicians of all nations to our great investigator."

Edmund Landau

Gösta Mittag-Leffler, Sweden, extended an invitation to the Congress to meet in Stockholm in 1916, which was accepted, on the motion of the President, by a voice vote.

Although the President's garden party had enjoyed the best weather of the week, he told his guests at the farewell banquet that he had found the rain "such that in a more superstitious age we should surely have concluded that heaven did not approve of our efforts; but fortunately today we regard it rather as a matter for the consideration of Section III (a) to decide why it is that solar radiation acting on a layer of compressible fluid on the planet should have selected England as the seat of its most unkindly efforts in the way of precipitation."

JULES HENRI POINCARÉ

"above praise"

April 29, 1854   **July 17, 1912**

Strasbourg 1920

Émile Picard

The summer following the signing of the Armistice the Inaugural General Assembly of the International Research Council (later described as "interalliée") met in Brussels to consider a series of post-war scientific congresses. The mathematicians present were not sufficiently numerous nor sufficiently accredited to form the proposed International Mathematical Union, but they decided to elect Charles de la Vallée-Poussin, Louvain, as their chairman and to hold an International Mathematical Congress in Strasbourg the following year.

"The above information of action taken six months ago now reaches American mathematicians for the first time in indirect ways," grumbled *The AMS Bulletin.* "Before the war international mathematical activities were carried on very efficiently by mathematicians, and it may be hoped that they will soon resume charge of their affairs."

In spite of objections to the idea, the Americans appointed L. E. Dickson, Chicago, chairman of a committee to consider an American section of the proposed IMU and later—with L. P. Eisenhart, Princeton—to represent the United States "at the meeting of the IMU" in Strasbourg.

Strasbourg lies in territory (Alsace-Lorraine) ceded by the French to the Germans in 1871 after the Franco-Prussian war and recently returned to the French. It was thus a particularly symbolic site for a congress which excluded mathematicians from the Central Powers and admitted those from neutral countries only after acceptance by a two-thirds vote of the executive committee. The whole idea of holding such a scientific meeting and calling it "an international congress of mathematicians" was opposed on several fronts (by G. H. Hardy and J. E. Littlewood in England, D. E. Smith in the United States, among others). Mittag-Leffler objected, not only that the congress was not international, but also that the next congress, resuming the series begun in Zürich, was, by previous agreement, to be held in Stockholm and *could not legally be held anywhere else.* Although he did not feel that he should go to Strasbourg himself, he was not—like Hardy and other "hotheads"—in favor of a boycott. He sent the assistant director of his institute, N. E. Nörlund, to make his point. At Nörlund's insistence, it was agreed to change the name of the Congress from "Congress of Mathematicians" to "Congress of Mathematics"; nevertheless, the *Comptes Rendus* bear the title "Congrès International des Mathématiciens." They do not, however, bear a number designating the congress as the sixth, and the question of recognizing Strasbourg as one in the series has been avoided by not numbering any subsequent congresses.

Camille Jordan

The International Mathematical Union was officially established at Strasbourg with de la Vallée-Poussin, who had been chairing the committee, as President. It was decided that in the future "les Congrès des mathématiques" would be held every four years under the auspices of the IMU and the International Research Council. The congress would be held in New York in 1924 and in Brussels in 1928.

Mathematicians from 27 countries attended the meeting, which was held from September 22 to 30 and presided over by Émile Picard, supported by Camille Jordan as Honorary President.

The large majority of talks (82 per cent) were, not unexpectedly, in French. English (17 per cent) was used by all the other speakers except Rudolf Fueter, Zürich, who delivered his talk in German. It appeared in that language in the *Comptes Rendus*. The first plenary address by Sir Joseph Larmor, Cambridge, it should be noted, treated in detail wartime work by Hilbert and Klein.

"Messieurs, le monde de 1920 est bien différent de celui du début de 1914," Picard said in the closing session of the Congress, "et il est peu d'hommes de science qui soient aujourd'hui disposés à s'isoler dans une tour d'ivoire; quoique savants, nous restons hommes. . . .

L. E. Dickson was sketched while giving his plenary address at Strasbourg.

**The Famous Astronomical
Clock of Strasbourg Cathedral**

"Quant à certaines relations, qui ont été rompues par la tragédie de ces dernières années, nos successeurs verront si un temps suffisamment long et un repentir sincère pourront permettre de les reprendre un jour, et si ceux qui se sont exclus du concert des nations civilisées sont dignes d'y rentrer. Pour nous, trop proches des événements, nous faisons encore nôtre la belle parole prononcée pendant la guerre par le cardinal Mercier, que, pardonner à certains crimes, c'est s'en faire le complice."

For those who may have missed the point, M. G. Koenigs, Paris, the Secretary-General, laid it out in the final words of the Congress: "en faisant de Strasbourg le siège du premier Congrès international d'après-guerre et en l'y organisant suivant le voeu de son coeur, ont eu sans aucun doute le désir complexe de donner:

à l'Alsace un témoignage de profonde affection,

à d'autres un exemple à suivre,

et à d'autres encore une leçon à méditer."

Among the Americans at Strasbourg, attending his first congress, was Norbert Wiener, who was to regret "my little share in sanctioning the meeting by my presence."

Toronto 1924

J. C. Fields

The second "congress of mathematics" was to take place in 1924 in New York City; but in April 1922 the Americans—for reasons not stated in their *Bulletin*—withdrew in favor of Toronto. The departure from a decision made at Strasbourg was construed as the quiet beginning of a revolt against the IMU.

Thanks to the dedication and industry of J. C. Fields, who crossed and re-crossed the Atlantic and traveled "thousands of kilometers" to make the arrangements and raise funds, the Congress took place on schedule but again without the mathematicians of the Central Powers.

Attendance at Toronto, while not at the level of pre-war congresses, was almost double that at Strasbourg.

The visiting mathematicians were most enthusiastically welcomed by the official representative of the Dominion:

"Those of us who, a quarter of a century ago, felt that, at not too distant a date, some development would take place, the effect of which could be to concentrate the world's attention on Canada, scarcely dreamed of such as this."

Fields, as President of the Congress, made clear to his guests that he and his countrymen were under no illusions about their present status in the mathematical world. Canadian mathematics had sprouted: "The tree is not yet large. May its growth be stimulated by this Congress!" Practically speaking, he hoped also that the Congress would not be without influence on the layman "to whom science must ultimately look for its material support." In Canada, a country of impressive mountains and rivers, the policy was "to accentuate more than has been done at previous Congresses the side of applied mathematics." Supplementing the traditional sections on applications, there would be a section on Electrical, Mechanical, Civil and Mining Engineering as well as one on Aeronautics, Naval Architecture, Ballistics and Radio Telegraphy.

Surprisingly, in view of the above, almost all the plenary speakers treated topics of pure mathematics. Speakers from abroad were Élie Cartan, Paris; Jean Marie Le Roux, Rennes; Salvatore Pincherle, Bologna;

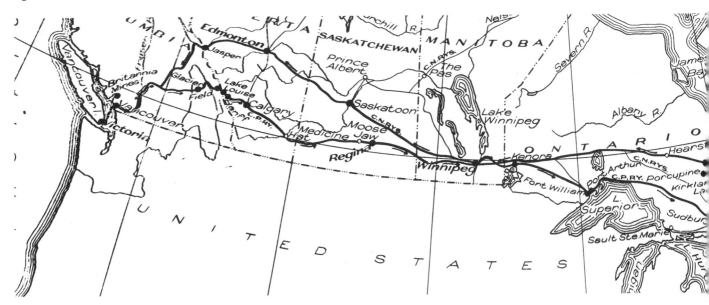

TO THE HEROES OF THE UNIVERSITY OF TORONTO
The Members of the International Mathematical Congress 1924
—Medallion inserted in the wall of the Memorial Tower

Francesco Severi, Rome; Carl Størmer, Oslo; and W. H. Young, the President of the London Mathematical Society. The latter managed to talk on "Some Characteristic Features of Twentieth Century Pure Mathematical Research" without mentioning Hilbert or any of his Paris Problems.

The bitterness still felt is evidenced in the opening paragraph of the speech of de la Vallée-Poussin in which he refers to the Strasbourg congress as more than a scientific congress: ". . . c'était un symbole et c'était une fête, celle de la délivrance de l'Alsace et aussi, comme je le disais alors, celle de la libération de la science que des mains sacrilèges avaient asservies trop longtemps à des dessins criminels."

Most American mathematicians, however, had always opposed the exclusion of mathematicians of Germany and the other Central Powers, and at Toronto they offered a resolution which was endorsed by Italy, Holland, Sweden, Denmark, Norway, and England:

> RESOLVED that the American Section of the International Mathematical Union requests the International Research Council to consider whether the time is ripe for the removal of restrictions on membership now imposed by the rules of the Council.

The Congress closed with money in its treasury, and Fields began to think about using it for an international mathematical prize.

"An Incident of Courtesy"
Charles de la Vallée-Poussin laying a wreath at the University of Toronto to honor the students lost in the First World War.

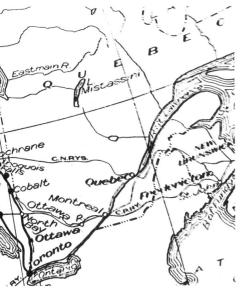

Route of the Roundtrip Transcontinental Excursion offered to members of the Congress and those attending the meeting of the British Association for the Advancement of Science.

W. H. Young

Bologna 1928

Salvatore Pincherle

The decision to hold the third post-war mathematical congress at the ancient university of Bologna was symbolic of a desire on the part of mathematicians to place their congresses again under scientific rather than governmental control. Although Germans and others who had been excluded at Strasbourg and Toronto were invited to attend, the olive branch, when it was finally offered, was not acceptable to all German mathematicians. The ensuing controversy over attending the Bologna congress revealed a division in their ranks which would only widen in coming years.

The leader of the group opposed to attending the Congress was Ludwig Bieberbach, Berlin. In the spring of 1928 he sent a letter to all German secondary schools and universities urging them to boycott the Congress. David Hilbert immediately responded with a letter of his own: "We are convinced that pursuing Herr Bieberbach's way will bring misfortune to German science and will expose us all to justifiable criticism from well disposed sides."

In August, although suffering from a recurrence of the illness which had been almost fatal a few years before, Hilbert personally led more than sixty German mathematicians to Bologna.

"It makes me very happy," he told the members of the Congress, "that after a long, hard time all mathematicians of the world are represented here. This is as it should be and as it must be for the prosperity of our beloved science.

"Let us consider that we as mathematicians stand on the highest pinnacle of the cultivation of the exact sciences. We have no other choice than to assume this highest place, because all limits, especially national ones, are contrary to the nature of mathematics. It is a complete misunderstanding of our science to construct differences according to people and races, and the reasons for which this has been done are very shabby ones.

"Mathematics knows no races . . . . For mathematics, the whole cultural world is a single country."

Unlike the earlier Italian congress, which had been held in the spring, the Bologna congress, organized by Salvatore Pincherle, took place during the first week of September so that more Americans could attend. G. D. Birkhoff, Harvard, who had been designated as chairman of the American delegation "if a chairman [was] necessary," delivered his remarks at the opening session in Italian and his later plenary speech in French. But he was no match for the Rector of the University, who welcomed the members of the Congress with an address in Latin.

David Hilbert

> . . . It was not mere chance that when the Nazis "purged"
> the German universities in 1933 their hand fell most
> heavily on the Hilbert school. . .
>
> — Hermann Weyl, 1944

The Palazzo Vecchio

G. D. Birkhoff

Hadamard at the Seaside near Ravenna

Among the plenary speakers was again Vito Volterra, to date the only mathematician to have delivered plenary addresses at four international congresses (1900, 1908, 1920, 1928). At Rome he had spoken in the presence of King Vittorio Emanuelle III. At Bologna, Benito Mussolini, a more powerful figure, had agreed to serve as the President of the Honorary Committee; but he did not put in an appearance.

The Congress was impressively peripatetic for a Congress, opening in Bologna, taking itself off to the beach at Ravenna, and closing in Florence. It was there, in the magnificent Palazzo Vecchio, that Birkhoff delivered his lecture on "Some Mathematical Aspects of Art."

Zurich 1932

Rudolf Fueter

The year 1932 marked the 35th anniversary of the First International Congress in Zürich. To many it apparently seemed an appropriate time to return to the traditionally neutral soil of Switzerland. According to the report in the published proceedings, Oswald Veblen, Princeton, the Chairman of the Delegation of the United States, "expressed the gratitude which all mathematicians must feel to Professors Fueter and Speiser for the generous way in which they came forward in Bologna, under very difficult circumstances [which he did not explain], and invited the Congress to Zürich."

Zürich thus became the first and the only city to host an international congress for a second time. As in 1897, participants gathered in the quarters of the Eidgenössische Technische Hochschule. Among those in attendance were twenty mathematicians who had been present in 1897, including C. F. Geiser, by then in his nineties, the President of the First International Congress. Also present was Henri Fehr of *L'Enseignement Mathématique,* the only mathematician who could say he had attended every one of the congresses to date.

President was Rudolf Fueter of the University of Zürich which, with the ETH, sponsored the meeting. Because of an economic crisis which was by then worldwide, the Congress had had to be organized "in a simple and dignified manner," Fueter explained in his opening address. Almost 700 mathematicians "of both sexes" (the first indication by a Congress president that some of the mathematicians in his audience might be women) had been able to make the journey to Zürich, although attendance was less by 169 than at Bologna. He hearkened back to the words of the late and "unforgettable" Hur-

witz which had in 1897 expressed so felicitously the need of mathematicians for personal intercourse with their fellows.

Organization of the Congress had been made somewhat difficult by the fact that the statutes of the International Mathematical Union had been allowed to expire in 1931. According to J. H. Weaver, writing at the time in *The American Mathematical Monthly,* the Union had been subjected to severe criticism because of its unsatisfactory work as well as its political nature: "At the insistence of American, English, and Danish mathematicians it was agreed to disband the Union and to ask the Congress to appoint a committee to investigate the need for a continuing body of some sort."

The return to internationality was also emphasized by the announcement of a bequest from the will of J. C. Fields, who had died earlier in the year, providing for two medals to be awarded at each congress — "of a character as purely international and impersonal as possible," as he had noted at an earlier time. The first medals were to be presented at the next congress in 1936.

Ludwig Bieberbach and Waclaw Sierpiński

<blockquote>
. . . to pay over the balance of the residue of my Estate. . . for the purpose of providing out of income thereof prizes to be attached to the International Mathematical Congress and also medals.

— Last will and testament of J. C. Fields, 1932
</blockquote>

In an overture to the German mathematicians who had opposed "die Bologna-gängers" in 1928, the organizing committee invited Ludwig Bieberbach to give one of the plenary addresses. The list of plenary speakers was again, as at Bologna, extensive and impressive. In a country which at that time did not allow women to vote, it was also distinguished by the inclusion of a woman mathematician—Emmy Noether of Göttingen.

Since the World Congress held in Chicago, women mathematicians had been in attendance at all but one of the international congresses. At Zürich in 1897, however, they were listed under "Damen," which classification consisted otherwise of the wives and daughters of the attending mathematicians. Attendance of women mathematicians through the second Zürich congress was as follows: Chicago, 3; Zürich, 4; Paris, 6; Heidelberg, 0; Rome, 19; Cambridge, 37; Strasbourg, 6; Toronto, 25; Bologna, 69; and Zürich, 35 (total congress attendance having declined). Because of the increasingly common custom of listing both men and women by initials, figures for many later congresses are not generally available. The only ascertainable facts are that the first of Emmy Noether has not been followed by a second, and that the number of women who have been invited to speak at international congresses since Noether does not differ very much from 0.

Emmy Noether, on the excursion on the Zürichsee.

Eidgenössische Technische Hochschule, c. 1925.

Oslo 1936

N. H. Abel
Maquette of the head
of the Vigeland statue.

By 1936, when the International Congress of Mathematicians convened in Oslo, Hitler had come to power in Germany and a number of the best known German mathematicians were already refugees from their homeland. In Italy, Mussolini had demanded an oath of loyalty from all university professors and, when Volterra had refused to sign, had dismissed him from the University of Rome and later from all Italian scientific academies. Although German supporters of the new regime and refugees were both at Oslo, not a single Italian male mathematician appeared. (Two women came.) The world's political and economic upheavals were also reflected in declining attendance: Bologna, 836; Zürich, 667; Oslo, 487.

The Congress, however, opened in style with a reception given by the King and Queen of Norway at the Royal Palace—a palace which has the distinction of having in its gardens a monumental sculpture of a mathematician. The statue by Vigeland depicts the young Abel triumphant over two vanquished figures which (according to Gabor Szegö) are the fifth degree polynomial equation and the elliptic function.

The name of Abel was also invoked by Carl Stormer, the President of the Congress, who in his opening address quoted the lines written by Bjornson for the Abel Jubilee in 1902:

"Impassible comme le temps/ est la science des nombres./ Leurs combinaisons sont/ dans une éternelle aurore/ plus pures que la neige,/ plus subtiles que l'air,/ mais plus fortes que le monde/ qu'elles pèsent sans balances/ qu'elles éclairent sans rayons."

Abel was not the only great figure in Norwegian mathematics to receive attention from the Congress. Members also took part in the dedication of a bust of Sophus Lie at the University of Oslo.

Halvdan Koht, Minister of Foreign Affairs, welcomed the mathematicians in words that must have surprised and gratified many:

"Indeed, I shall not blame you for having been of practical and even economic utility to the rest of mankind," he said. "But I prefer to lay stress upon another aspect of your studies. Though not, myself, belonging to the initiates, I venture to praise your science as leading in the expansion of the human mind."

The King, who was also present at the opening session, made the first award of the recently endowed "Fields Medals" to Lars Ahlfors, 29, Harvard, and Jesse Douglas, 39, Massachusetts Institute of Technology (whose medal in his absence was accepted by Norbert Wiener). A description of the work for which the medals had been awarded was given by Constantin Carathéodory,

Élie Cartan

Munich, who with Élie Cartan, Paris, was on his way to Harvard to represent European mathematics at that university's tercentenary celebration. At Oslo, for the third time before an international congress (1924-1932-1936), Cartan delivered a plenary address.

There had not been a mathematical congress in the United States since the "World Congress" of 1893, when that country had been still without a mathematical tradition of its own. By the time of Oslo, as Oswald Veblen, who headed the newly founded Institute for Advanced Study, later recalled, "important discoveries had been made by American mathematicians. New branches of mathematics were being cultivated, and new tendencies in research were showing themselves. Some American universities were receiving research workers from overseas, and interchanges of all sorts tended to be more and more on terms of equality. At the same time mathematics had attained a small but growing amount of recognition from the rest of the American community—enough at least to encourage us to invite the mathematicians of the world to a congress in this country in 1940."

The invitation was extended by George David Birkhoff and accepted by the Congress.

Constantin Carathéodory

Dinner aboard the Stavangerfjord
The excursion on Oslo Fjord was on the Norwegian transatlantic steamer Stavangerfjord. Dinner was served in four dining rooms. On the left can be seen Veblen and Fréchet, on the right Wiener and Weyl.

Cambridge 1950

Oswald Veblen

When the mathematicians of almost all the world gathered for their first post-war congress in Cambridge, Massachusetts, in the summer of 1950, "American mathematics" had experienced a metamorphosis.

The "colonial period" had ended by the time of the Oslo congress.

"Now, fourteen years have elapsed . . ., and we are approaching the end of another epoch, . . . the period during which North America has absorbed so many powerful mathematicians from all over the world that the indigenous traditions and tendencies of mathematical thought have been radically changed as well as enriched," pointed out Congress President Oswald Veblen, who had succeeded G. D. Birkhoff, since the latter's death in 1944, as the leader of the American mathematicians.

"These American gains have seemed to be at the cost of great losses to European mathematics," Veblen conceded. "But there are so many signs of vitality in Europe that it is now possible to hope that the losses will be only temporary while the American gains will be permanent."

From the time that the Second World War began, one year before the scheduled 1940 congress, leaders of the American Mathematical Society had been determined that there should be no congress until such time that the gathering could be truly international. It had unfortunately not been possible to achieve this goal in an era when the post-war superpowers, so recently allies, battled one another in "the cold war." The Cambridge Congress, as Veblen pointed out, was meeting "in the shadow of another crisis, perhaps even more menacing than that of 1940. . . ."

Officers of the Congress had tried to obtain a visa for every mathematician who wished to attend, J. R. Kline, the Secretary, reported: "As far as they know only one mathematician from any independent nation was prevented from attending the Congress because he failed to pass a political test and this man did not notify the officers of the Congress about his difficulties. Only two

Salomon Bochner, Princeton, born in Poland.

Shiing-Shen Chern, Chicago, born in China.

Kurt Gödel, IAS, born in Austria.

mathematicians from occupied countries failed to secure visas. Mathematicians from behind the Iron Curtain were uniformly prevented from attending the Congress by their own governments."

Before the opening session, however, a cablegram was received from S. Vavilov, the President of the USSR Academy of Sciences, wishing the Congress success.

In spite of the absence of the Eastern bloc, the actual attendance (1700) was more than twice that of the largest previous congress. President Veblen was to be only the first post-war president to complain about the size of the congress he presided over.

"Mathematics is terribly individual," he told the members. "Any mathematical act, whether of creation or apprehension, takes place in the deepest recesses of the individual mind. Mathematical thoughts must nevertheless be communicated. . . . But the ideal communication is to a very few other individuals. By the time it becomes necessary to raise one's voice in a large hall some of the best mathematicians I know are simply horrified and remain silent."

Nevertheless, 22 mathematicians, 15 of them affiliated with American institutions if not American born, raised their voices to deliver invited addresses. Their names were as impressive as any listed at earlier congresses, but they were new names. An exception was that of "Cartan," now carried by Henri Cartan, the son of Élie Cartan.

Fields Medals went to Laurent Schwartz, France, and Atle Selberg, United States (although Norwegian-born).

Harald Bohr

In describing Selberg's share in the elementary proof of the prime number theorem, Harald Bohr quoted a remark of the late G. H. Hardy: "If anyone produces an elementary proof of the prime number theorem, he will show that . . . the subject does not hang together in the way we have supposed, and that it is time for the books to be cast aside and for the theory to be rewritten."

Shizuo Kakutani, Yale, born in Japan.

John Von Neumann, IAS, born in Hungary.

Abraham Wald, Columbia, born in Rumania.

Amsterdam 1954

were from the University of California at Berkeley. Another foreign-born American speaker was John von Neumann, who had been invited to lay before the Congress some great problems, in the spirit of Hilbert, which would make for comparable progress in the latter half of the century. Von Neumann spoke on "Unsolved Problems" but, overworked and already mortally ill, he never submitted his manuscript for publication.

J. A. Schouten

"The series of the International Congresses of which this is to be one are very loosely held together," Oswald Veblen told the 1553 mathematicians gathered in the world famous Concertgebouw of Amsterdam during the first week of September 1954. ". . . To symbolize the tenuous continuity . . ., the President of the old Congress emerges for a moment from the obscurity in which he belongs, to propose the name of the person selected by the hosts of the new congress to preside over it."

The name that Veblen proposed was that of J. A. Schouten, President of the Mathematical Society of the Netherlands, who moments before had welcomed the members of the Congress in seven languages: Dutch, English, French, German, Italian, Russian, and Swedish.

In his presidential address Schouten dwelled on the change in the role of mathematics following the Second World War: "During and after the war it became obvious to everyone that nearly all branches of modern society in war and in peace need a lot of mathematics of all kinds." His point was underlined by the plenary address of a countryman, David van Dantzig, who spoke of mathematical problems raised by the floods of the previous year which had devastated the Netherlands.

The Organizing Committee had put together an all-star cast of speakers for the plenary sessions, including four mathematicians from Moscow (P. S. Alexandrov, I. M. Gelfand, A. N. Kolmogorov, S. M. Nikolski) and one from Warsaw (Karol Borsuk). There were other Russian and Polish born mathematicians among the speakers, but these—Jerzy Neyman and Alfred Tarski—

IN MEMORIAM

1870 HENRI FEHR 1954

Founder, with C. A. Laisant, of *L'Enseignement Mathématique*

Henri Fehr, Geneva, was among the mathematicians attending the First International Congress in Zürich in 1897. Subsequently he attended and generally took an active part in every congress up to and including Cambridge in 1950. He died in the congress year of 1954.

> [These] are not congresses of mathematics, that highly
> organized body of knowledge, but of mathematicians, those
> rather chaotic individuals who create and conserve it.
> — Oswald Veblen, 1954

A. N. Kolmogorov

The continuing increase in attendance at international congresses again disturbed its president: ". . . there is a limit to congresses of this kind. This limit will perhaps be reached very soon if the number of mathematicians goes on increasing as rapidly as it does now. . . ."

Highlight of the Congress was a lecture by the old Hermann Weyl in which he presented the mathematical work of the Fields Medalists, Kunihiko Kodaira and Jean-Pierre Serre, both more than thirty years younger, in a manner which showed him, as a friend and colleague wrote after Weyl's death the following year, "well abreast of those new theories which have changed the face of mathematics in the last twenty years."

In 1936 Carathéodory as Chairman of the Selection Committee for the Fields Medal had presented the work of both recipients to the Congress, and in 1950 the Chairman, Harald Bohr, had done the same. After Weyl's virtuoso performance, however, it became customary for a specialist in the field to take over the presentation of each individual medalist's work.

1885 THE LAST CONGRESS OF HERMANN WEYL 1955

Excerpted from Weyl's address as President of the Fields Medal Committee:

. . . I hope the Congress as a whole will approve our choice. In justification of it let me say this: by study and information we became convinced that Serre and Kodaira had not only made highly original and important contributions to mathematics in recent years, but that these hold out great promise for future fruitful non-analytic (will say: non-foreseeable) continuation.

. . . I realize how difficult it is for a man of my age to keep abreast of the rapid development . . . which the young generation forces upon our old science. . . . [The burden] rests more heavily on my than on my predecessors' shoulders; for while they reported on things within the circle of classical analysis, where every mathematician is at home, I must speak on achievements that have a less familiar conceptual basis. . . . Be prepared then to have to listen now to a short lecture on cohomology, linear differential forms, faisceaux or sheaves, Kähler manifolds and complex line bundles. . . .

. . . If I omitted essential parts or misrepresented others, I ask your pardon, Dr. Serre and Dr. Kodaira; it is not easy for an older man to follow your striding paces.

. . . The mathematical community is proud of the work you both have done. It shows that the old gnarled tree of mathematics is still full of sap and life. Carry on as you began!

Edinburgh 1958

Sir Edmund Whittaker

In the summer of 1958 mathematicians gathered for their third post-war congress in Edinburgh, the romantic stronghold of the Scotch. It had been hoped that the Congress would be presided over by Sir Edmund Whittaker, "that great figure in the life of the City and University of Edinburgh, so much respected and loved in the mathematical world." Whittaker, however, had died in 1956; and W. V. D. Hodge, Cambridge, regretfully took his place.

The increasing size of the congress (now 1658 regular members) continued to cause concern. The Secretary, in describing the preparations, which had involved almost all the institutions and services of the City of Edinburgh, resorted on occasion to the word "onerous." President Hodge openly pondered the question whether, with the increasing number of international symposia in special fields, some of them even sponsored by the IMU, international congresses of mathematicians were still necessary. He concluded that the answer was yes.

"It is essential for the well-being of mathematics that there should be periodic gatherings attended by representatives of all branches of the subject, and this for several reasons: in my personal opinion, the most important reason is that gatherings such as this serve as an invaluable safeguard against the dangers of excessive specialization."

Although youth had not been expressly stipulated in the terms of the gift from Fields, Heinz Hopf, Chairman of the Selection Committee for the Fields Medals, explained that the Committee had decided to continue the tradition of choosing young mathematicians as the re-

cipients of the medals: "However, the other day, a friend of mine made the remark that when one looks at the present situation in mathematics and the development of recent years, one feels that it is the old rather than the young who need encouragement."

Hopf described the achievements of René Thom, France; and Harold Davenport, those of Klaus Roth, Great Britain. Quoting the remark of Lewis Carroll's Duchess that there is a moral in everything if only one can find it, Davenport said, "It is not difficult to find a moral in Dr. Roth's work. It is that the great unsolved problems of mathematics may still yield to direct attack, however

Edinburgh University, founded in 1583.

Four Logicians on a Boat
P. S. Novikov, S. Kleene, R. L. Goodstein, and
A. Church on the steamship excursion on the Clyde.

John Napier (1550-1617)
Edinburgh's most celebrated
mathematician.

Heinz Hopf

difficult and forbidding they appear to be, and however
much effort has already been spent on them."

There was a significant innovation at Edinburgh. Since
the Congress in 1897, Algebra and Number Theory—
what Hilbert had called at Heidelberg "the grammar of
our science"—had been Section I. At Edinburgh it was
preceded by Logic and the Foundations of Mathematics.

Non-mathematical activities gave the visitors an idea
of Scottish folkways, history, and contemporary culture.
Hosts for international congresses have almost invariably
planned some excursion by water, and the Edinburghers
did not fail in this respect, offering a steamer cruise from
Glasgow down the Clyde.

The Congress adjourned with the question of where it
would next meet still unsettled: "I am authorized to say,"
President Hodge announced, ". . . that while for reasons
of a technical nature it is not possible to make any announcement today of the name of the host country for
1962, the prospects of holding a Congress in that year
amount to a certainty."

Stockholm 1962

Rolf Nevanlinna

In 1962 the fourth post-war International Congress of Mathematicians met in Stockholm, a full half century after the mathematicians in Cambridge, England, had accepted Mittag-Leffler's invitation to hold their next congress in his country.

The Swedes had hesitated at Edinburgh about assuming the responsibility of a congress. The decision had been a very hard one, as Otto Frostman, the Chairman of the Organizing Committee, conceded at the opening session. In addition to the concern about the large numbers who would be attending, there was also a more important concern: "the development of mathematics itself which is proceeding so rapidly that no man can survey but a part or parts of the front line, and total coverage can only be achieved by joint work on an international basis."

In order to present "a scientific program worthy of an international congress," the Swedish committee had worked closely with a small Consultative Committee from the International Mathematical Union. The result-

ing plenary program—which for the first time since the war included two German mathematicians—did not include a single Swedish mathematician; but the Swedes were very pleased with the use of a consultative committee and recommended it to future congresses. In view of the assistance thus given by the IMU, they felt it appropriate that their congress be presided over, not by a Swede, but by the Finnish mathematician Rolf Nevanlinna, who, in addition to being President of the Congress, was also President of the International Mathematical Union and Chairman of the Fields Medal Committee—a triple role which has not been duplicated by any other congress president.

As Chairman of the Fields Medal Committee, Nevanlinna presented Lars Hörmander, Sweden, and John W. Milnor, United States, to the King of Sweden, who awarded them their medals. Hörmander's work was described to the Congress by another Swede, Lars Gårding, also a specialist in partial differential equations. A fellow American, Hassler Whitney, describing Milnor's work, recalled how in the 1930's, concerned with the relations between differential and topological properties, he had felt that his own contributions were merely beginnings of what might come in the future. Now he could say: "The future has arrived!"

Whatever their doubts about their ability to organize a scientific program, the Swedes could have had no doubts about being able to provide a rich cultural program. This included the Swedish Ballet at the Royal Opera House, Pergolesi's *Il maestro di musica* at the 18th century court

Hassler Whitney

> The development of mathematics would soon lead to an impossible situation, were not another tendency working against it: the tendency towards synthesis.
>
> — Rolf Nevanlinna, 1962

Gösta Mittag-Leffler

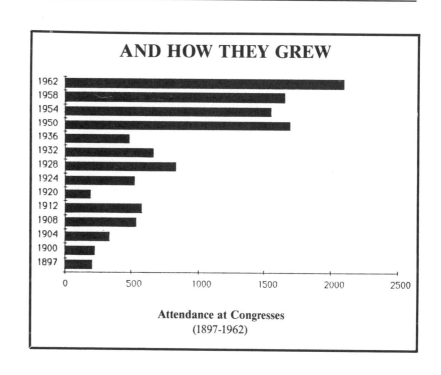

AND HOW THEY GREW

Attendance at Congresses
(1897-1962)

theatre at Drottningholm, a concert of chamber music in the Konserthuset in Stockholm, and a recital by the Swedish tenor, Nicolai Gedda, at the park at Skansen — all in addition to the obligatory congress excursion on water — a cruise to the outer Archipelago.

At the closing session of the Congress, M. A. Lavrentiev, USSR, issued an invitation in Russian to the Congress to meet in 1966 in Moscow. The invitation was translated into English by P. S. Alexandrov and accepted — in French — by President Nevanlinna.

Royal Institute of Technology

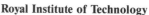

Whitney, in the 1930's.

Moscow 1966

Even the endowment of the Fields Medals had been bolstered by an anonymous gift so that—Georges de Rham, Chairman of the Selection Committee, explained—it had appeared that, in view of the expansion of mathematics since the first medals were awarded 30 years earlier, the number "could judiciously be increased to four." Even then, the Committee had had to give up some names: "Several young mathematicians of extraordinary brilliance were among them. But because they are so young, there will be many congresses before they reach forty . . ." The chosen four—Michael Atiyah, Great Britain; Paul J. Cohen, United States; Alexander Grothendieck, France; and Stephen Smale, United States—received their medals from Academician M. V. Keldysh, the President of the Academy of Sciences.

Alonzo Church, Princeton, who presented Cohen's work to the Congress, felt that—despite Cohen's contributions to several other fields—he should devote his whole time to his solving—"better, the unsolving," he said (quoting W. W. Boone)—of the problem of the continuum hypothesis, which Hilbert had placed first on his famous list.

"The feeling that there is an absolute realm of sets, somehow determined in spite of the non-existence of a complete axiomatic characterization, receives more of a blow [from Cohen's work] than from the famous Gödel incompleteness theorems," Church told the Congress. " . . . The Gödel-Cohen results and subsequent extensions of them have the consequence that there is not one set theory but many, with the difference arising in connection with a problem which intuition still seems to tell us must 'really' have only one true solution."

Moscow University

Alonzo Church

The fifth post-war international congress of mathematicians in Moscow—a heady mixture of mathematics, vodka and caviar, operas, ballets and museums, and the ever present politics of East and West—opened in the enormous Palace of Congresses, a Soviet addition to the sprawling Kremlin of the Russian Czars. The size of the hall was appropriate, for the congress was to be additionally distinguished in almost every aspect of its program by size.

Of the 5,594 persons who had registered, over four thousand were in attendance—twice as many as had attended any earlier congress—a record which has stood for two decades. Forty-nine countries were represented, the greatest number up to that time; and the 15 sections just barely missed being twice the number at the preceding congress.

After the opening session, the Congress adjourned to the skyscraper Moscow University which dominates the skyline of the city. Plenary speeches were given there, and section meetings convened. The choice of speakers at the Congress was explained by its president, I. G. Petrovski, the Rector of the University, as follows:

"All decisions of the International Mathematical Union regarding the number of sections, the choices of hour and half-hour addresses were fully accepted by the Soviet Organizational Committee. We only allowed ourselves to add several talks to these recommendations."

There were 17 plenary speakers: 9 from Great Britain and the United States together, 5 from Russia, 2 from West Germany, and 1 from France. As if by agreement, each spoke in the language of his own country.

The Fields Medalist Stephen Smale had been described by René Thom as "un pionnier qui prend ses risques avec un courage tranquille." During the Congress, he made headlines when he held a press conference on the steps of Moscow University criticizing the US for its involvement in Vietnam but also the USSR for the lack in that country of "the basic means of protest."

ICM Postage Stamp
issued on the occasion of the Moscow Congress of 1966.

P. S. Alexandrov, L. E. J. Brouwer, P. S. Urysohn
Alexandrov and Urysohn, pictured here in the early 1920's, were the first Russian mathematicians to go abroad to study after the revolution of 1917.

Nice 1970

Paul Montel

which, since Cambridge, had crowded congress schedules, 265 printed *communications individuelles* were distributed. When Jean Dieudonné, Chairman of the Organization Committee, asked whether the members approved the continuation of this organization, the response was unanimously in favor.

The fact that Dieudonné posed his question in English as well as in French was indicative of the change that had occurred in the languages used at congresses. Although the program was printed in French, English and German, English has become the *lingua franca.* All the plenary speakers, including the Russians, gave their talks in that language with the exception of Pontrjagin, who used French. The Russians included I. M. Gelfand, addressing his third international congress (1954-1962-1970) as a plenary speaker.

Although attendance at Nice (2811) was considerably below that at Moscow, the number of countries represented (60) exceeded Moscow by 11. And, once again, four Fields Medals, instead of the previously traditional two, were awarded. The names of the Medalists were announced in alphabetical order by Henri Cartan, Chairman of the Selection Committee: Alan Baker, Great Britain; Heisuke Hironaka, Japan; Sergei Novikov, Soviet Union; and John G. Thompson, United States. Novikov had not been able to attend the Congress — the second time in history a Medalist had not received his medal in person. The following year, however, when the Executive Council of the IMU was being entertained at a luncheon in Moscow, Cartan personally awarded the Medal to Novikov.

The sixth post-war international congress of mathematicians, held in Nice at the beginning of September 1970, was the third to take place in France — a number not to be equaled by any other country until this year when, at Berkeley, the United States hosts its third congress. The 94-year-old Paul Montel, as Honorary President, joined the company of Hermite and Jordan, who had been so honored at earlier French congresses.

Under Jean Leray, President, the Congress was organized in an efficient and innovative manner. Each morning, at plenary sessions, the members heard two one-hour *conférences générales.* Each afternoon they were offered *un choix d'exposés spécialisés,* limited to 50 minutes each so that they might hear as many as three in an afternoon. In place of the 10-minute oral presentations of papers

Jean Dieudonné

The new Medalists and all French mathematicians who had previously been awarded Fields Medals were received in Paris by the President of the Republic, Georges Pompidou. It was the custom, the newpaper *Nice-Matin* noted — more patriotically than accurately, never to award the *medaille Fields* to a mathematician of the country in which the Congress was being held.

Paul Turán, Hungary, pointed out to the Congress that two important points were exemplified by the work of one of the Fields Medalists, Alan Baker: "Firstly, that beside the worthy tendency to start a theory in order to solve a problem it pays also to attack specific difficult problems directly. . . . Secondly, it shows that a direct solution of a deep problem develops itself quite naturally into a healthy theory and gets into early and fruitful contact with other significant problems of mathematics. So," he concluded, "let the two different ways of doing mathematics live in peaceful coexistence for the benefit of our science."

The beginning of the congress year had seen the solution by a 22-year-old Russian of one of Hilbert's most intractable problems: *To find a general method of solution for diophantine equations* (the tenth problem). Yuri Matijasevic̃, Leningrad, who had showed that such a method as Hilbert required *does not exist,* talked on his work before Section I on Logic and the Foundations of Mathematics.

Yuri Matijasevic̃

Henri Cartan

LANGUAGES USED BY PLENARY SPEAKERS

		English	French	German	Italian	Russian
1897	Zürich	—	1	2	1	—
1900	Paris	—	4	—	—	—
1904	Heidelberg	1	1	1	1	—
1908	Rome	1	6	1	2	—
1912	Cambridge, U.K.	5	1	1	1	—
1920	Strasbourg	2	3	—	—	—
1924	Toronto	4	3	—	1	—
1928	Bologna	2	5	3	6	—
1932	Zürich	2	9	10	—	—
1936	Oslo	7	2	10	—	—
1950	Cambridge, U.S.A.	19	2	1	—	—
1954	Amsterdam	14	3	—	—	3
1958	Edinburgh	13	3	2	—	1
1962	Stockholm	9	3	1	—	3
1966	Moscow	9	1	2	—	5
1970	Nice	15	1	—	—	—
1974	Vancouver	15	2	—	—	—
1978	Helsinki	15	—	—	—	—
1983	Warsaw	13	—	—	—	—

Vancouver 1974

Totem Pole Emblem, official congress insignia.

The second international mathematical congress in Canada took place on the opposite side of the North American continent from the first. The city of Vancouver had been the last stop on the great trans-continental train trip offered to mathematicians attending the congress in Toronto in 1924. Half a century later more than eight times as many mathematicians gathered on the scenic campus of the University of British Columbia for their seventh congress since the end of World War II.

In his opening remarks, however, President H. S. M. Coxeter, Toronto, did not turn back to the words of J. C. Fields, the President of the 1924 Congress, but to those of J. A. Schouten, who had presided over the Congress in Amsterdam in 1954. At that time Schouten had pointed out that, following the war, the place of mathematics in the world had changed entirely.

"What he meant," it seemed to Coxeter, "was that, whereas formerly mathematics was studied by exceptional people, in ivory towers, the subject had become immensely popular. Even sport was affected: footballs (for soccer) began to be made to look like truncated icosahedra, electronic computers were springing up everywhere, and departments of mathematics in all universities were expanding to accommodate crowds of eager students."

The explosion of mathematical activity, Coxeter had found, could be exactly ascertained with a tape measure applied to successive 11-year periods of *Mathematical Reviews,* each of which was double that which preceded it. He foresaw an end to the doubling in the near future. The present generation seemed essentially anti-intellectual. Pure mathematics was being abandoned everywhere in favor of the applications. The change in emphasis had been so great that jobs for young researchers were almost non-existent.

"What then should be our advice to a student who is wondering whether to specialize in mathematics? In view of the present scarcity of suitable jobs, I would advise him to take up some other subject, unless his love for mathematics is so intense that he finds himself doing it in almost all his spare time, even thinking about it while sleeping, between dreams."

The number of Fields Medals returned to two at Vancouver. They were awarded to Enrico Bombieri, Italy, and David Mumford, United States. Komarauolu Chandrasekharan described Bombieri's work; and John Tate, that of Mumford.

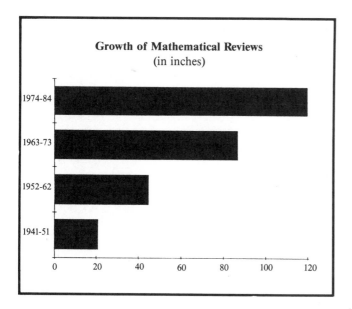

Growth of Mathematical Reviews
(in inches)

David Mumford, Harvard, receiving Fields Medal from K. Chandrasekharan (center), Chairman of the Fields Committee, with H. S. M. Coxeter, President of the Congress, looking on.

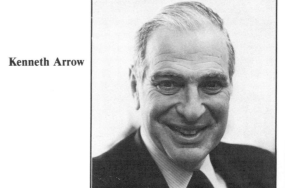

Kenneth Arrow

Nobel Prizes and Mathematics

The widely circulated explanation of why there are no Nobel Prizes in mathematics is that Alfred Nobel wanted to make sure that the mathematician Mittag-Leffler would never be awarded a Nobel Prize. By 1974, with increasing mathematization of all the sciences, it had begun to seem that one day a mathematician might indeed become a Nobelist, albeit in a field other than mathematics. Kenneth Arrow had already (in 1972) won the Prize for highly mathematical work in economics. Eleven years later the Nobel Prize for economics would go to Gerard Debreu, one of the plenary speakers at Vancouver. Debreu holds a joint appointment at Berkeley in economics — and mathematics.

Gerard Debreu

Helsinki 1978

Rolf Nevanlinna and **Olli Lehto**

The International Congress of Mathematicians in the capital of Finland during the summer of 1978 was presided over by Olli Lehto, Helsinki, who was supported by the revered Finnish mathematician Rolf Nevanlinna as Honorary President. With more than three thousand registered members accompanied by another thousand or so associate members, from 83 countries, it was the largest scientific gathering ever to take place in Finland.

There were 15 plenary addresses in addition to more than two thousand individual papers. The first plenary speaker had been, in 1936, the first man to receive a Fields Medal (having an alphabetical advantage over the other medalist). Lars Ahlfors, after many years on the faculty at Harvard, returned to the city of his birth to deliver for the third time a plenary address before an inter-

national congress (1936-1962-1978). For the third time in history four Fields Medals were awarded. Recipients were Jacques Deligne, France; Charles Fefferman, United States; G. A. Margulis, USSR (who was not present at the Congress); and Daniel Quillen, United States. The plenary group also included three young mathematicians who in four years time would all receive Fields Medals: Alain Connes, William P. Thurston, and S.-T. Yau.

Finlandia Hall, seating 1700, was filled to overflowing when André Weil, IAS, delivered his lecture on "The History of Mathematics: Why and How," which drew 2500 auditors, even with a misprint regarding the time. Fortunately the situation had been anticipated, and television monitors outside the hall permitted those who could not get in to see and hear Weil. It was also Weil's third plenary address before an international congress (1950-1954-1978).

The Helsinki Congress was further distinguished when Yurii I. Manin, Academy of Sciences, Moscow, became the first invited speaker to take as his subject the emblem of a congress. He urged the members of his audience to look carefully at it.

"You will easily recognize the design. This is part of the famous 'modular configuration' consisting of (a) the Lobachevsky plane modeled on the complex unit disk; (b) the set of fundamental triangles for the theta-group bounded by Lobachevsky lines. To get the whole picture, draw the central triangle and perform consecutive reflections relative to its sides."

In spite of the richness of the scientific program and the cultural and scenic pleasures of the Finnish capital, the Congress was marred by a lack of cooperation on the part of the Soviet Union. Writing carefully after the close of the Congress, Armand Borel, the Chairman of the

Congress Emblem

> Thus my original question, "Why mathematical history?" finally reduces to the question "Why mathematics?" which fortunately I do not feel called upon to answer.
> —André Weil, 1978

Armand Borel

Consultative Committee, which had been entrusted since 1974 with all matters pertaining to the scientific program, concentrated on the most important point of disagreement—the selection of invited speakers.

"In a number of cases, the initial opinions within the [Committee] were far apart and a lot of discussion was needed. It was gratifying that, eventually, except in two or three cases, all the decisions were unanimous. I showed the list of (mostly Soviet) invited speakers at the Vancouver Congress who had accepted but [had been] unable to come, and expressed the hope that most invited speakers would accept and, this time, be able to attend the Congress . . . Professor Nikolskii stated that he would do all he could to help the invited Soviet mathematicians to come . . . However, in April, . . . Professor Pontrjagin stated in conversations that the list of invited Soviet mathematicians was unpopular within the Soviet National Committee. When it was pointed out to him that practically all the decisions of the C.C. had been unanimous, he answered that Professor Nikolskii had worked as an individual and that the point of view of the Soviet National Committee had not been correctly presented. To my knowledge, there was never any further official statement on the matter . . . Eventually, fifteen (out of thirty) Soviet invited speakers were not able to come to the Congress (and the total Soviet membership at the Congress was ridiculously small) . . .

"Whatever [the] causes, such actions are of course detrimental to the ICM and, besides, render any agreement on procedures meaningless. In fact, if they become more widespread, they would even make the ICM itself meaningless."

André Weil

Warsaw 1982/3

← oops, huh?!

For the first time in the peace-time history of international mathematical congresses, five years elapsed between the congresses.

A heated debate—first over holding a congress and later over attending—divided much of the mathematical community throughout 1982 and the first half of 1983, particularly in the United States. On one side were those who felt that going to Warsaw would indicate support for the Polish government; on the other, those who felt that going would indicate support for those who opposed the government. According to G. D. Mostow, the Chairman of the U.S. National Committee for Mathematics, the General Assembly of the International Mathematical Union, debating whether to go ahead with plans to hold the Congress, was greatly moved by a two-sentence statement (at the top of the following page) by the Polish mathematician Andrzej Schinzel.

Against this background of turmoil and uncertainty, Czesław Olech and his committee, with the active support of the Polish government, struggled valiantly to bring an international congress into being on Polish soil. At the opening session at the Palace of Culture in August 1983, Olech was chosen President of the Congress; and Władysław Orlicz, whom he described as "the Nestor of Polish mathematicians," was named Honorary President.

Aleksander Gieysztor, the President of the Polish Academy of Sciences, delivering a welcoming address, reminded his audience that during the Second World War well over half the actively working Polish mathematicians had lost their lives and many others had had to leave their native land. Universities, libraries and printing presses had been largely destroyed, the education system wiped out, and scientific activity disrupted: "The fact that this Congress is being held in Warsaw in 1983, 38 years after the war, gives evidence of the reconstruction of Polish science both in the organization and the substantive sense. In particular, it is a proof of the renaissance and expansion of the Polish mathematical community."

Gieysztor's avoidance of contemporary political references was "much appreciated," according to Mostow, "by those participants anxious to avoid tacit approval of martial law imprisonments." Some French, American, and English speakers dedicated their talks to Polish mathematicians who had been interned or imprisoned.

The scientific program for the Congress was again the responsibility of the IMU's Consultative Committee.

"[But] the final scientific result," Olech pointed out in his presidential address, "depends not only on those who have prepared the program but also on the cooperation of those who have been chosen to fill in the program with the invited survey lectures, both plenary and in sections. I regret that you will not have the opportunity to listen to some of the lectures announced . . . even in the printed program you have just received."

Despite the fact that a number of western mathematicians declined to attend and some of those who had registered did not appear, attendance turned out to be respectably high: 2300 as compared to 3000 at the previous Helsinki congress, and the number of countries represented exceeded by one the number at Helsinki. The United States delegation, although it had dropped from 600 to 100, was nevertheless the largest contingent after those of Poland and the USSR. Congress participants also included 33 young mathematicians from 21 developing countries who were able to attend as a result of travel grants from the IMU's Special Development Fund.

Robert Tarjan

The Committee for the Fields Medals, headed by Lennart Carleson, had announced its selections in the summer of 1982. Now, however, Professor Orlicz presented the medals themselves to Alain Connes, France; William Thurston, United States; and Shing-Tung Yau, United States. The prizewinning achievements were enthusiastically described by Hozihiro Araki (of Connes, "[he has led the theory of operator algebras] to breathtaking achievements beyond the expectation of experts"); by C. T. C. Wall (of Thurston, "his ideas have completely revolutionized the study of topology in 2 and 3 dimensions, and brought about a new and fruitful interplay between analysis, topology and geometry"); and by Louis Nirenberg (of Yau, "an analyst's geometer (or geometer's

> [As to the moral aspect of the question], for almost two thousand years it has been held a good and charitable deed to visit people in prison. The condition of the martial law being rather hard, this may apply here.
>
> —Andrzej Schinzel, 1982

analyst) . . . He has succeeded in solving problems on which progress has been stopped for years"). Since neither Wall nor Nirenberg was in attendance, their talks were read to the audience. Also awarded at Warsaw, and also previously announced by a committee headed by J. L. Lions, was the first Nevanlinna Prize, a tribute by Finland to her great mathematician, Rolf Nevanlinna, for his contributions to science in general and to Finnish computer science in particular. The first award went to Robert Tarjan, United States, for outstanding contributions to mathematical aspects of information science.

"Pure mathematics enjoys the luxury of studying its constructions, whether finite or infinite, in complete independence of all questions of efficiency," explained Jacob Schwartz, who spoke on Tarjan's work. "By contrast, theoretical computer science must ultimately concern itself with computing engines which operate with limited speed and data storage, and therefore must take efficiency as one of its central concerns. Two closely related activities, algorithm design and algorithm analysis, grow out of this inevitable concern . . . Robert Tarjan has been a leader in both these enterprises, which lie at the intellectual heart of computer science . . ."

To give their visitors a glimpse of traditional Polish folk life, the Poles arranged, among other activities, two picnic excursions to Bogusławice. These featured a show called "The Cracovian Wedding"—described in the *Proceedings* of the Congress as "a spectacle comprising folk songs and dances in colorful dresses and a parade of riders and equipages driven by horses in the traditional Cracovian harnesses." Throughout the Congress the visitors were able to travel free on the trams and busses of Warsaw.

"It is of course not up to me to make a general evaluation in public of whether our decision [to hold ICM-82 in Warsaw in 1983] was correct or not," Olli Lehto, the Secretary of the International Mathematical Union, said at the closing session of the Congress. "But let me say that I feel very happy that the ICM-82 took place here. The opportunity of international cooperation was maintained, and in spite of regrettable absences of some invited speakers, this was a high class meeting from a scientific point of view." He concluded by emphasizing the basic principle of the IMU, that politics should never find a foothold within the Union: "As individuals, we may of course have whatever political views we choose, but when it amounts to organized international cooperation in mathematics, then political aspects should be put aside entirely. Our fine science should be the uniting link between us and make us in a true sense one big mathematical family."

Congress Chronology

August 1978	Poland's invitation to hold the 1982 International Congress of Mathematicians in Warsaw is accepted by acclamation at Helsinki.
Spring 1980	Polish industrial workers begin to strike locally.
August 1980	Strikes spread throughout Poland. Lech Walesa emerges as leader of Inter-factory Strike Committee.
February 1981	Jaruzelski replaces Pinkowski as Premier in major reconstruction of Polish government.
July 1981	First Announcement of Warsaw Congress is mailed to mathematical community.
September 1981	Solidarity's 9.5 million members hold their first congress and elect Lech Walesa their leader.
December 1981	Second Announcement of Warsaw Congress is mailed. Martial Law is declared in Poland. Lech Walesa is arrested.
January 1982	United States Government blocks use of federal travel funds for scientific meetings in Warsaw in spite of recommendations against the block by scientists in government and the National Academy of Sciences.
April 1982	International Mathematical Union postpones Warsaw Congress.
August 1982	IMU, meeting in Warsaw, with 79 participants from 37 countries, decides to hold the Warsaw Congress in 1983.
November 1982	IMU's Executive Committee confirms the decision.
August 1983	International Congress of Mathematicians take place in Warsaw as rescheduled.

Berkeley 1986

Andrew M. Gleason

Since the "World Congress of Mathematicians," held in Chicago in 1893, the mathematicians of the world — as urged then by Felix Klein — have gone far in forming unions and holding international congresses. In the summer of 1986 the twentieth such congress took place at the University of California (Berkeley).

The city of Berkeley from which the University takes its designation was named for the Anglican bishop, George Berkeley (1685-1753), not for his perceptive comments regarding the newly invented calculus, but for another perceptive comment — "Westward the course of empire takes its way" — which occurs in a work entitled "On the Prospect of Planting Arts and Learning in America."

Under its President, Andrew Gleason, the Congress opened at an early morning session in the outdoor Greek Theater of the University. On the fiftieth anniversary of the awarding of the first Fields Medals, it was voted by acclamation that the Honorary President should be Lars Ahlfors, the first mathematician to receive one of the coveted gold medals. Ahlfors personally presented Fields Medals to Simon Donaldson (Great Britain), Gerd Faltings (West Germany) and Michael Freedman (United States) and the Nevanlinna Prize to Leslie Valiant (Great Britain).

At a subsequent session John Milnor and Michael Atiyah in turn described the respective achievements of Freedman and Donaldson, both of whom have made important contributions to the understanding of 4-dimensional space. Barry Mazur, who described Falting's proof of the 60-year-old Mordell conjecture, called it "one of the great moments in mathematics." Volker Strassen, speaking on the work of the recipient of the Nevanlinna Prize, stressed the fact that Valiant has contributed in a decisive way to the growth of almost every branch of the fast growing young tree of theoretical computer science, his theory of *counting problems* being perhaps his most important and mature work.

The Congress itself was distinguished by an increasing emphasis on computer science. *The New York Times* headlined MATHEMATICIANS FINALLY LOG ON. Steve Smale, Berkeley's own Fields Medalist (Moscow 1966), led off the stellar lineup of fifteen plenary speakers with a lecture on "Complexity aspects of numerical analysis" — a far cry from his Moscow lecture on "Differentiable dynamical systems." The problem-solving abilities of computers, he pointed out, have created a challenge that is philosophical, logical and mathematical: "This subject is now likely to change mathematics itself. Algorithms become a subject of study, not just a means of solving problems."

Thanks to computer graphics, the Congress was also much more visual and colorful than those which preceded it. When, however, reporters pounced on the prizewinners with the question that they most like to ask mathematicians, all three Fields Medalists responded that computers were of no use in their work. Valiant, the specialist in theoretical computer science, conceded that no, he didn't use a computer either.

East-West political frictions continued. The papers of almost half the invited speakers from the Soviet Union had to be read in absentia, and the USSR itself was represented at less than half-strength. Jürgen Moser, the president of the IMU, expressed the hope of everyone present that the next congress (in Kyoto in 1990) would be truly international.

Leslie Valiant

> [Mathematical research is] a very personal and individual process. It might be defined as the difference between what you know and what everyone else knows.
>
> —Michael Freedman, 1986

The distaff side of the mathematical family was as usual not very well represented; and Marina Ratner, a mathematics professor at Berkeley, circulated a statement explaining that she was boycotting the Congress: "Since the founding of the International Congress its leadership has conducted a *de facto* 'female free' policy. . . . Now I wish to address the conscience of the mathematical community. Why has this situation been tolerated for so many years?"

The future of mathematics was also much on the minds of many who attended the Congress, particularly those from the United States. Richard Johnson, Acting Scientific Adviser to President Reagan, urged the mathematicians to concentrate on mathematics education. John Addison, Chairman of the Berkeley Mathematics Department, spoke to the press about the imminent retirement of mathematicians trained during the post-sputnik "boom" and the fact that business and industry are draining off a fair number of the comparatively few research mathematicians now being trained by the universities. And Michael Freedman said of his Fields Medal that it carried with it "the responsibility to nurture mathematics. . . . Ultimately trying to move society in the direction where elementary school children grow up liking mathematics instead of hating it. . . . The country is facing a crisis in terms of comprehension and liking of mathematics in contrast to rising Pacific Rim countries." The auditorium in which the plenary lectures were delivered was also the site of daily sessions of the International Commission on Mathematical Instruction.

In accordance with custom, there were a number of excursions to local scenic attractions. In addition, not to be outdone in the matter of national folkways, the organizers of the Congress arranged a western barbecue and rodeo for their guests. Those interested in visiting the quarters of the local mathematics department found them in Evans Hall. There, in the mathematics library, they saw a portrait of Griffith C. Evans (1887-1973), who is honored at Berkeley for transforming its once modest mathematics department into one of the world's great centers of mathematical research. Next to it is a portrait of Evans' predecessor, M. W. Haskell (1873-1948), one of the many American students of Felix Klein.

Award Winners: Valiant, Freedman, Faltings, Donaldson

Fields Medalists

The International Medals for Outstanding Discoveries in Mathematics, commonly known as the Fields Medals, do not bear the name or the likeness of J. C. Fields. The head appearing on the medals, which are minted in gold by the Royal Canadian Mint, is that of Archimedes. Though the terms of the endowment do not mention any restriction as to the age of the recipients, the medals have traditionally been awarded to mathematicians under 40 to comply with Fields' wish that they be an encouragement to further work by the recipients.

J. L. Synge, whom Fields named as his executor, was active in making the arrangements for the setting up of the awards, which were first made at the International Congress of 1936 in Oslo. Because of the intervention of World War II, there were no further awards until 1950. On the following pages all the Medalists to date are listed alphabetically with year of award and affiliation at time of award.

The Opening Session of the Oslo Congress.
Three of the members of the Fields Selection Committee are seen in the front row—second, third and fourth from the left are E. Cartan, Fueter, and Carathéodory. Behind them on the aisle is Ahlfors and next to him, partially hidden, is Wiener, who accepted the Fields Medal for Douglas. King Haakon VII is seen in the aisle two-thirds of the way back in the hall.

John Charles Fields (1863-1932)

J. C. Fields, a professor at the University of Toronto, organized the international congress held in Toronto in 1924 and served as its president. He was so successful in obtaining financial support that the Congress ended with a surplus. This provided the initial funding for the Fields Medals. Later these were further endowed from his estate.

Fields was born in Hamilton, Ontario. He received his A.B. in 1884 from the University of Toronto, being awarded, perhaps significantly for future mathematicians, a gold medal for his work. Three years later he took his Ph.D. at Johns Hopkins. After a brief period on the faculty of Pennsylvania's Allegheny College, he went to Germany and studied with Fuchs, Frobenius, Hensel and H. A. Schwarz. It was at this time that he developed his lasting friendship with Mittag-Leffler. In 1902 he returned to the University of Toronto and remained there until his death thirty years later. He was responsible for some significant mathematics in his time and received a number of honors, including election to the Royal Society; but today he is remembered for the medals which, contrary to his personal wish, are known by his name.

For more information on Fields, see Henry S. Tropp, "The Origins and History of the Fields Medals," *Historia Mathematica* 3 (1976), 167-181.

**1936
AHLFORS, Lars Valerian
b. April 18, 1907, Helsinki
Harvard University**

**1966
ATIYAH, Michael Francis
b. April 22, 1929, London
Oxford University**

Awarded medal for research on covering surfaces related to Riemann surfaces of inverse functions of entire and meromorphic functions. Opened up a new field of analysis.

Did joint work with Hirzebruch in *K*-theory; proved jointly with Singer the index theorem of elliptic operators on complex manifolds; worked in collaboration with Bott to prove a fixed point theorem related to the "Lefschetz formula."

Fields Medalists

1974
BOMBIERI, Enrico
b. November 26, 1940
Milan
University of Pisa

1970
BAKER, Alan
b. August 19, 1939, London
Cambridge University

Generalized the Gelfond-Schneider theorem (the solution to Hilbert's seventh problem). From this work he generated transcendental numbers not previously identified.

Major contributions in the distribution of primes, in univalent functions and the local Bieberbach conjecture, in theory of functions of several complex variables, and in theory of partial differential equations and minimal surfaces — in particular, to the solution of Bernstein's problem in higher dimensions.

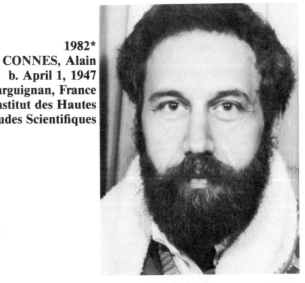

1982*
CONNES, Alain
b. April 1, 1947
Darguignan, France
Institut des Hautes
Études Scientifiques

1966
COHEN, Paul Joseph
b. April 2, 1934
Long Branch, New Jersey
Stanford University

Used technique called "forcing" to prove the independence in set theory of the axiom of choice and of the generalized continuum hypothesis. The latter problem was the first of Hilbert's problems of the 1900 Congress.

Contributed to the theory of operator algebras, particularly the general classification and a structure theorem for factors of type III, classification of automorphisms of the hyperfinite factor, classification of injective factors, and applications of the theory of C*-algebras to foliations and differential geometry in general.

Fields Medalists

1978
DELIGNE, Pierre René
b. October 3, 1944
Brussels, Belgium
Institut des Hautes
Études Scientifiques

1986
DONALDSON, Simon K.
b. August 20, 1957
Cambridge, England
Oxford University

Gave solution of the three Weil conjectures concerning generalizations of the Riemann hypothesis to finite fields. His work did much to unify algebraic geometry and algebraic number theory.

Received medal primarily for his work on topology of four-manifolds, especially for showing that there is a differential structure on euclidean four-space which is different from the usual structure.

1936
DOUGLAS, Jesse
b. July 3, 1897
New York City
Massachusetts Institute
of Technology

1986
FALTINGS, Gerd
b. July 28, 1954
Gelsenkirchen-Buer
West Germany
Princeton University

Did important work on the Plateau problem which is concerned with finding minimal surfaces connecting and determined by some fixed boundary.

Using methods of arithmetic algebraic geometry, he received medal primarily for his proof of the Mordell Conjecture.

Fields Medalists

1986
FREEDMAN, Michael H.
b. April 21, 1951, Los Angeles
University of California, San Diego

1978
FEFFERMAN, Charles Louis
b. April 18, 1949
Washington, D.C.
Princeton University

Contributed several innovations that revised the study of multidimensional complex analysis by finding correct generalizations of classical (low-dimensional) results.

Developed new methods for topological analysis of four-manifolds. One of his results is a proof of the four-dimensional Poincaré Conjecture.

1970
HIRONAKA, Heisuke
b. April 9, 1931
Yamaguchi-Ken, Japan
Harvard University

1966
GROTHENDIECK, Alexander
b. March 28, 1928, Berlin
University of Paris

Built on work of Weil and Zariski and effected fundamental advances in algebraic geometry. He introduced the idea of K-theory (the Grothendieck groups and rings). Revolutionized homological algebra in his celebrated "Tohoku paper."

Generalized work of Zariski who had proved for dimension $\leqslant 3$ the theorem concerning the resolution of singularities on an algebraic variety. Hironaka proved the result in any dimension.

Fields Medalists

1962
HÖRMANDER, Lars
b. January 24, 1931
Mjällby/Blekinge, Sweden
Stockholm University

1954
KODAIRA, Kunihiko
b. March 16, 1915
Tokyo
Princeton University

Worked in partial differential equations. Specifically, contributed to the general theory of linear differential operators. The questions go back to one of Hilbert's problems at the 1900 Congress.

Achieved major results in the theory of harmonic integrals and numerous applications to Kählerian and more specifically to algebraic varieties. He demonstrated, by sheaf cohomology, that such varieties are Hodge manifolds.

1978
MARGULIS, Gregori Aleksandrovitch
b. February 24, 1946, Moscow
Moscow University

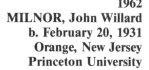

1962
MILNOR, John Willard
b. February 20, 1931
Orange, New Jersey
Princeton University

Provided innovative analysis of the structure of Lie groups. His work belongs to combinatorics, differential geometry, ergodic theory, dynamical systems, and Lie groups.

Proved that a 7-dimensional sphere can have several differential structures; this led to the creation of the field of differential topology.

Fields Medalists

1970
NOVIKOV, Serge
b. March 20, 1938
Gorki, USSR
Belorusskii University

1974
MUMFORD, David Bryant
b. June 11, 1937
Worth, Sussex, England
Harvard University

Contributed to problems of the existence and structure of varieties of moduli, varieties whose points parametrize isomorphism classes of some type of geometric object. Also made several important contributions to the theory of algebraic surfaces.

Made important advances in topology, the most well-known being his proof of the topological invariance of the Pontrjagin classes of a differentiable manifold. His work included a study of the cohomology and homotopy of Thom spaces.

1958
ROTH, Klaus Friedrich
b. October 29, 1925
Breslau, Germany
University of London

1978
QUILLEN, Daniel G.
b. June 27, 1940
Orange, New Jersey
Massachusetts Institute
of Technology

The prime architect of the higher algebraic K-theory, a new tool that successfully employed geometric and topological methods and ideas to formulate and solve major problems in algebra, particularly ring theory and module theory.

Solved in 1955 the famous Thue-Siegel problem concerning the approximation to algebraic numbers by rational numbers and proved in 1952 that a sequence with no three numbers in arithmetic progression has zero density (a conjecture of Erdös and Turán of 1935).

Fields Medalists

1950
SCHWARTZ, Laurent
b. March 5, 1915, Paris
University of Nancy

1950
SELBERG, Atle
b. June 14, 1917
Langesund, Norway
Institute for Advanced Study

Developed the theory of distributions, a new notion of generalized function motivated by the Dirac delta-function of theoretical physics.

Developed generalizations of the sieve methods of Viggo Brun; achieved major results on zeros of the Riemann zeta function; gave an elementary proof of the prime number theorem (with P. Erdös), with a generalization to prime numbers in an arbitrary arithmetic progression.

1954
SERRE, Jean-Pierre
b. September 15, 1926
Bages, France
Collège de France

1966
SMALE, Stephen
b. July 15, 1930
Flint, Michigan
University of California,
Berkeley

Achieved major results on the homotopy groups of spheres, especially in his use of the method of spectral sequences. Reformulated and extended some of the main results of complex variable theory in terms of sheaves.

Worked in differential topology where he proved the generalized Poincaré conjecture in dimension $n \geq 5$: Every closed, n-dimensional manifold homotopy-equivalent to the n-dimensional sphere is homeomorphic to it. Introduced the method of handle-bodies to solve this and related problems.

Fields Medalists

1970
THOMPSON, John Griggs
b. October 13, 1932, Kansas
University of Chicago

1958
THOM, René
b. September 2, 1923
Montbeliard, France
University of Strasbourg

In 1954 invented and developed the theory of cobordism in algebraic topology. This classification of manifolds used homotopy theory in a fundamental way and became a prime example of a general cohomology theory.

Proved jointly with W. Feit that all non-cyclic finite simple groups have even order. The extension of this work by Thompson determined the minimal simple finite groups, that is, the simple finite groups whose proper subgroups are solvable.

1982*
YAU, Shing-Tung
b. April 4, 1949
Kwuntung, China
Institute for Advanced Study

1982*
THURSTON, William P.
b. October 30, 1946
Washington, D.C.
Princeton University

Revolutionized study of topology in 2 and 3 dimensions, showing interplay between analysis, topology, and geometry. Contributed idea that a very large class of closed 3-manifolds carry a hyperbolic structure.

*The awards were announced in 1982 even though the Warsaw Congress was not held until 1983.

Made contributions in differential equations, also to the Calabi conjecture in algebraic geometry, to the positive mass conjecture of general relativity theory, and to real and complex Monge-Ampère equations.

Plenary Lectures

It has always been customary to invite a number of outstanding mathematicians to deliver lectures before plenary sessions of international congresses. These invited lectures have gone under different names at different congresses: for consistency they are all listed here as Plenary Lectures. An asterisk following a speaker's name indicates that he was also a plenary speaker at a later congress. His second plenary address is indicated by two asterisks, and so on. An (R) following a title indicates that the talk was originally given in Russian. Because of typographical limitations all Russian titles have had to be translated into English or German.

CHICAGO (August 21-26, 1893)

Klein, Felix*: The Present State of Mathematics.

ZÜRICH (August 9-11, 1897)

Hurwitz, Adolf: Über die Entwickelung der allgemeinen Theorie der analytischen Funktionen in neuerer Zeit.

Klein, Felix**: Zur Frage des höheren mathematischen Unterrichtes.

Peano, Giuseppe: Logica matematica.

Poincaré, Henri*: Sur les rapports de l'analyse pure et de la physique mathématique.

PARIS (August 6-12, 1900)

Cantor, Moritz: L'historiographie des mathématiques.

Mittag-Leffler, Gösta*: Une page de la vie de Weierstrass.

Poincaré, Henri**: Du rôle de l'intuition et de la logiqúe en mathématiques.

Volterra, Vito*: Betti, Brioschi, Casorati—Trois analystes italiens et trois manières d'envisager les questions d'analyse.

HEIDELBERG (August 8-13, 1904)

Greenhill, Alfred George: The Mathematical Theory of the Top Considered Historically.

Painlevé, Paul: Le problème moderne de l'intégration des équations différentielles.

Segre, Corrado: La geometria d'oggidì e i suoi legami coll'analisi.

Wirtinger, Wilhelm: Riemanns Vorlesungen über die hypergeometrische Reihe und ihre Bedeutung.

ROME (April 6-11, 1908)

Darboux, Gaston: Les origines, les méthodes et les problèmes de la géométrie infinitésimale.

Dyck, Walther v.: Die Encyklopädie der mathematischen Wissenschaften.

Forsyth, Andrew Russell: On the Present Condition of Partial Differential Equations of the Second Order as Regards Formal Integration.

Lorentz, Hendrik Antoon: Le partage de l'énergie entre la matière pondérable et l'éther.

Mittag-Leffler, Gösta**: Sur la représentation arithmétique des fonctions analytiques générales d'une variable complexe.

Newcomb, Simon: La théorie du mouvement de la lune: son histoire et son état actuel.

Picard, Émile: La mathématique dans ses rapports avec la physique.

Poincaré, Henri***: L'avenir des mathématiques.

Veronese, Giuseppe: La geometria non-archimedea.

Volterra, Vito**: Le matematiche in Italia nella seconda metà del secolo XIX.

Plenary Lectures

CAMBRIDGE, UK (August 22-28, 1912)

Bôcher, Maxime: Boundary Problems in One Dimension.

Borel, Émile*: Définition et domaine d'existence des fonctions monogènes uniformes.

Brown, Ernest W.: Periodicities in the Solar System.

Enriques, Federigo: Il significato della critica dei principii nello sviluppo delle matematiche.

Galitzen, Prince B.: The Principles of Instrumental Seismology.

Landau, Edmund: Gelöste und ungelöste Probleme aus der Theorie der Primzahlverteilung und der Riemannschen Zetafunktion.

Larmor, Joseph*: On the Dynamics of Radiation.

White, W. H.: The Place of Mathematics in Engineering Practice.

STRASBOURG (September 22-30, 1920)

Dickson, Leonard Eugene*: Some Relations between the Theory of Numbers and Other Branches of Mathematics.

Larmor, Joseph**: Questions in Physical Interdetermination.

Nörlund, Niels Erik: Sur les équations aux différences finies.

Vallée-Poussin, Charles de la: Sur les fonctions à variation bornée et les questions qui s'y rattachent.

Volterra, Vito***: Sur l'enseignement de la physique mathématique et de quelques points d'analyse.

TORONTO (August 11-16, 1924)

Cartan, Élie*: La théorie des groupes et les recherches récentes de géométrie différentielle.

Dickson, Leonard Eugene**: Outline of the Theory to Date of the Arithmetics of Algebras.

Le Roux, Jean Marie: Considérations sur une équation aux dérivées partielles de la physique mathématique.

Pierpont, James: Non-Euclidean Geometry from Non-Projective Standpoint.

Pincherle, Salvatore: Sulle operazioni funzionali lineari.

Severi, Francesco*: La géométrie algébrique.

Størmer, Carl*: Modern Norwegian Researches on the Aurora Borealis.

Young, William H.*: Some Characteristic Features of Twentieth Century Pure Mathematical Research.

BOLOGNA (September 3-10, 1928)

Amoroso, Luigi: Le equazioni differenziali della dinamica economica.

Birkhoff, George David*: Quelques éléments mathématiques de l'art.

Borel, Émile**: Le calcul des probabilités et les sciences exactes.

Castelnuovo, Guido: La geometria algebrica e la scuola italiana.

Fréchet, Maurice*: L'analyse générale et les espaces abstraits.

Hadamard, Jacques: Le développement et le rôle scientifique du calcul fonctionnel.

Hilbert, David: Probleme der Grundlegung der Mathematik.

Kármán, Theodore von: Mathematische Probleme der modernen Aerodynamik.

Lusin, Nikolai N.: Sur les voies de le théorie des ensembles.

Marcolongo, Roberto: Leonardo da Vinci nella storia della matematica e della meccanica.

Plenary Lectures

Puppini, Umberto: Le bonifiche in Italia.

Tonelli, Leonida: Il contributo italiano alla teoria delle funzioni di variabili reali.

Veblen, Oswald*: Differential Invariants and Geometry.

Volterra, Vito****: La teoria dei funzionali applicata ai fenomeni ereditari.

Weyl, Hermann: Kontinuierliche Gruppen und ihre Darstellungen durch lineare Transformationen.

Young, William H.**: The Mathematical Method and Its Limitations.

ZÜRICH (September 5-12, 1932)

Alexander, James W.: Some Problems in Topology.

Bernstein, Serge: Sur les liaisons entre quantités aléatoires.

Bieberbach, Ludwig: Operationsbereiche von Funktionen.

Bohr, Harald: Fastperiodische Funktionen einer komplexen Veränderlichen.

Carathéodory, Constantin: Über die analytischen Abbildungen durch Funktionen mehrerer Veränderlicher.

Carleman, Torsten: Sur la théorie des équations intégrales linéaires et ses applications.

Cartan, Élie**: Sur les espaces riemanniens symétriques.

Fueter, Rudolf*: Idealtheorie und Funktionentheorie.

Julia, Gaston: Essai sur le développement de la théorie des fonctions de variables complexes.

Menger, Karl: Neuere Methoden und Probleme der Geometrie.

Morse, Marston*: The Calculus of Variations in the Large.

Nevanlinna, Rolf: Über die Riemannsche Fläche einer analytischen Funktion.

Noether, Emmy: Hyperkomplexe Systeme in ihren Beziehungen zur kommutativen Algebra und zur Zahlentheorie.

Pauli, Wolfgang: Mathematische Methoden der Quantenmechanik.

Riesz, Frédéric: Sur l'existence de la dérivée des fonctions d'une variable réelle et des fonctions d'intervalle.

Severi, Francesco**: La théorie générale des fonctions analytiques de plusieurs variables et la géométrie algébrique.

Sierpiński, Wacław: Sur les ensembles de points qu'on sait définir effectivement.

Stenzel, J.: Anschauung und Denken in der klassischen Theorie der griechischen Mathematik.

Tschebotaröw, N.: Die Aufgaben der modernen Galoisschen Theorie.

Valiron, Georges: Le théorème de Borel-Julia dans la théorie des fonctions méromorphes.

Wavre, Rolin: L'aspect analytique du problème des figures planétaires.

OSLO (July 13-18, 1936)

Ahlfors, Lars V.*: Geometrie der Riemannschen Flächen.

Banach, Stefan: Die Theorie der Operationen und ihre Bedeutung für die Analysis.

Birkhoff, George David**: On the Foundations of Quantum Mechanics.

Bjerknes, Vilhelm: New Lines in Hydrodynamics.

Cartan, Élie***: Quelques aperçus sur le rôle de la théorie des groupes de Sophus Lie dans le développement de la géométrie moderne.

Corput, Jan G. van der: Diophantische Approximationen.

Plenary Lectures

Fréchet, Maurice**: Mélanges mathématiques.

Fueter, Rudolf**: Die Theorie der regulären Funktionen einer Quaternionenvariablen.

Hasse, Helmut: Über die Riemannsche Vermutung in Funktionenkörpern.

Hecke, Erich: Neuere Fortschritte in der Theorie der elliptischen Modulfunktionen.

Mordell, Louis Joel: Minkowski's Theorems and Hypotheses on Linear Forms.

Neugebauer, Otto: Über vorgriechische Mathematik und ihre Stellung zur griechischen.

Nielsen, Jakob: Topologie der Flächenabbildungen.

Ore, Øystein: The Decomposition Theorems of Algebra.

Oseen, C. W.: Probleme der geometrischen Optik.

Siegel, Carl Ludwig: Analytische Theorie der quadratischen Formen.

Størmer, Carl**: Programme for the Quantitative Discussion of Electron Orbits in the Field of a Magnetic Dipole, with Application to Cosmic Rays and Kindred Phenomena.

Veblen, Oswald**: Spinors and Projective Geometry.

Wiener, Norbert*: Gap Theorems.

CAMBRIDGE, USA (August 30-September 6, 1950)

Albert, A. Adrian: Power-Associative Algebras.

Beurling, Arne: On Null-Sets in Harmonic Analysis and Function Theory.

Bochner, Salomon: Laplace Operator on Manifolds.

Cartan, Henri*: Problèmes globaux dans la théorie des fonctions analytiques de plusieurs variables complexes.

Chern, Shiing-shen*: Differential Geometry of Fiber Bundles.

Davenport, Harold: Recent Progress in the Geometry of Numbers.

Gödel, Kurt: Rotating Universes in General Relativity Theory.

Hodge, W. V. D.: The Topological Invariants of Algebraic Varieties.

Hopf, Heinz: Die n-dimensionalen Sphären und projektiven Räume in der Topologie.

Hurewicz, Witold: Homology and Homotopy.

Kakutani, Shizuo: Ergodic Theory.

Morse, Marston**: Recent Advances in Variational Theory in the Large.

Neumann, John von*: Shock Interaction and Its Mathematical Aspects.

Ritt, Joseph Fels: Differential Groups.

Rome, Adolfe: The Calculation of an Eclipse of the Sun According to Theon of Alexandria.

Schwartz, Laurent: Théorie des Noyaux.

Wald, Abraham: Basic Ideas of a General Theory of Statistical Decision Rules.

Weil, André*: Number Theory and Algebraic Geometry.

Whitney, Hassler: r-Dimensional Integration in n-Space.

Wiener, Norbert**: Comprehensive View of Prediction Theory.

Wilder, Raymond L.: The Cultural Basis of Mathematics.

Zariski, Oscar: The Fundamental Ideas of Abstract Algebraic Geometry.

Plenary Lectures

AMSTERDAM (September 2-9, 1954)

Alexandrov, P. S.: Aus der mengentheoretischen Topologie der letzten zwanzig Jahren. (R)

Borsuk, Karol: Sur l'élimination de phénomènes paradoxaux en topologie générale.

Brauer, Richard: On the Structure of Groups of Finite Order.

Dantzig, David van: Mathematical Problems Raised by the Flood Disaster 1953.

Dieudonné, Jean: Le calcul différentiel dans les corps de caractéristique p > 0.

Gelfand, I. M.*: Some Aspects of Functional Analysis and Algebra.

Goldstein, S.: On Some Methods of Approximation in Fluid Mechanics.

Harish-Chandra*: Representations of Semisimple Lie Groups.

Jessen, Borge: Some Aspects of the Theory of Almost Periodic Functions.

Kolmogorov, A. N.: Théorie générale des systèmes dynamiques et mécanique classique. (R)

Lichnerovicz, André: Les groupes d'holonomie et leurs applications.

Neumann, John von**: On Unsolved Problems in Mathematics.

Neyman, Jerzy: Current Problems of Mathematical Statistics.

Nikolskii, S. M.: Einige Fragen der Approximation von Funktionen durch Polynome. (R)

Segre, Beniamino: Geometry upon an Algebraic Variety.

Stiefel, Edward Ludwig: Recent Developments in Relaxation Techniques.

Tarski, Alfred: Mathematics and Metamathematics.

Titchmarsh, Edward C.: Eigenfunction Problems Arising from Differential Equations.

Weil, André**: Abstract versus Classical Algebraic Geometry.

Yosida, Kosaku: Semi-Group Theory and the Integration Problem of Diffusion Equations.

EDINBURGH (August 14-21, 1958)

Alexandrov, A. D.: Modern Development of Surface Theory.

Bogolyubov, N. N., and Vladimirov, V. S.: On Some Mathematical Problems of Quantum Field Theory.

Cartan, Henri**: Sur les fonctions de plusieurs variables complexes: les espaces analytiques.

Chevalley, Claude: La théorie des groupes algébriques.

Eilenberg, Samuel: Applications of Homological Algebra in Topology.

Feller, William: Some New Connections between Probability and Classical Analysis.

Gårding, Lars: Some Trends and Problems in Linear Partial Differential Equations.

Grothendieck, Alexander: The Cohomology Theory of Abstract Algebraic Varieties.

Hirzebruch, Friedrich: Komplexe Mannigfaltigkeiten.

Kleene, Stephen C.: Mathematical Logic: Constructive and Non-Constructive Operations.

Lanczos, Cornelius: Extended Boundary Value Problems.

Pontrjagin, L. S.*: Optimal Processes of Regulation. (R)

Roth, Klaus F.: Rational Approximations to Algebraic Numbers.

Schiffer, Menaham M.: Extremum Problems and Variational Methods in Conformal Mapping.

Steenrod, Norman E.: Cohomology Operations and Symmetric Products.

Plenary Lectures

Temple, George: Linearization and Delinearization.

Thom, René: Des variétés triangulées aux variétés différentiables.

Uhlenbeck, George Eugene: Some Fundamental Problems in Statistical Physics.

Wielandt, Helmut W.: Entwicklungslinien in der Strukturtheorie der endlichen Gruppen.

STOCKHOLM (August 15-22, 1962)

Ahlfors, Lars V.**: Teichmüller Spaces.

Borel, Armand: Arithmetic Properties of Linear Algebraic Groups.

Church, Alonzo: Logic, Arithmetic, and Automata.

Dynkin, E. B.: Markov Processes and Problems in Analysis. (R)

Eckmann, Beno: Homotopy and Cohomology Theory.

Gelfand, I. M.**: Automorphic Functions and the Theory of Representations. (R)

Grauert, Hans: Die Bedeutung des Levischen Problems für die analytische und algebraische Geometrie.

Henrici, Peter: Problems of Stability and Error Propagation in the Numerical Integration of Ordinary Differential Equations.

Kahane, Jean-Pierre: Transformées de Fourier des fonctions sommables.

Milnor, John W.: Topological Manifolds and Smooth Manifolds.

Newman, M. H. A.: Geometrical Topology.

Nirenberg, Louis: Some Aspects of Linear and Nonlinear Partial Differential Equations.

Šafarevič, I. R.: Algebraic Number Fields. (R)

Selberg, Atle: Discontinuous Groups and Harmonic Analysis.

Serre, Jean-Pierre: Géométrie algébrique.

Tits, Jacques*: Groupes simples et géométries associées.

MOSCOW (August 16-26, 1966)

Adams, John F.: A Survey of Homotopy-Theory.

Artin, Michael: The Etale Topology of Schemes.

Atiyah, Michael F.: Global Aspects of the Theory of Elliptic Differential Operators.

Bellman, Richard: Dynamic Programming and Modern Control Theory.

Carleson, Lennart: Convergence and Summability of Fourier Series.

Efimov, N. V.: Hyperbolic Problems in the Theory of Surfaces. (R)

Harish-Chandra**: Harmonic Analysis on Semisimple Lie Groups.

Krein, M. G.: Analytic Problems and Results in the Theory of Linear Operators in Hilbert Space. (R)

Malgrange, Bernard: Théorie Locale des Fonctions Différentiables.

Malcev, A. I.: On Some Questions on the Border of Algebra and Logic. (R)

Piatetski-Shapiro, I. I.: Automorphic Functions and Arithmetic Groups. (R)

Schröder, Johann: Ungleichungen und Fehlerabschätzungen.

Schütte, Kurt: Neuere Ergebnisse der Beweistheorie.

Plenary Lectures

Smale, Stephen*: Differentiable Dynamical Systems.

Stein, Charles M.: Some Recent Developments in Mathematical Statistics.

Thompson, John G.: Characterizations of Finite Simple Groups.

Vinogradov, I. M., and Postnikov, A. G.: Recent Developments in Analytic Number Theory. (R)

NICE (September 1-10, 1970)

Baker, Alan: Effective Methods in the Theory of Numbers.

Bott, Raoul: On Topological Obstructions to Integrability.

Browder, William: Manifolds and Homotopy Theory.

Chern, Shiing-shen**: Differential Geometry: Its Past and Its Future.

Feit, Walter: The Current Situation in the Theory of Finite Simple Groups.

Gelfand, I. M.***: The Cohomology of Infinite Dimensional Lie Algebras; Some Questions of Integral Geometry.

Griffiths, Phillip A.: A Transcendental Method in Algebraic Geometry.

Hörmander, Lars: Linear Differential Operators.

Kato, Tosio: Scattering Theory and Perturbation of Continuous Spectra.

Keisler, H. Jerome: Model Theory.

Marchŭk, G. I.: Methods and Problems of Computational Mathematics.

Pontrjagin, L. S**: Les Jeux différentiels linéaires.

Stein, Elias M.*: Some Problems in Harmonic Analysis Suggested by Symmetric Spaces and Semi-Simple Groups.

Swan, Richard G.: Algebraic K-Theory.

Tate, John: Symbols in Arithmetic.

Wall, C. T. C.: Geometric Topology: Manifolds and Structures.

VANCOUVER (August 21-29, 1974)

Arnold V. I.*: Critical Points of Smooth Functions.

Bauer, Heinz: Aspects of Modern Potential Theory.

Bombieri, Enrico: Variational Problems and Elliptic Equations.

Debreu, Gerard: Four Aspects of the Mathematical Theory of Economic Equilibrium.

Deligne, Pierre: Poids dans la cohomologie des variétés algébriques.

Duff, George F. D.: Mathematical Problems of Tidal Energy.

Fefferman, Charles: Recent Progress in Classical Fourier Analysis.

Glimm, James: Analysis over Infinite-Dimensional Spaces and Applications to Quantum Field Theory.

Kreiss, Heinz-Otto: Initial Boundary Value Problems for Hyperbolic Partial Differential Equations.

Lions, Jacques L.: Sur la théorie du controle.

Milner, Eric C.: Transversal Theory.

Quillen, Daniel: Higher Algebraic K-Theory.

Schmidt, Wolfgang M.: Applications of Thue's Method in Various Branches of Number Theory.

Plenary Lectures

Singer, I. M.: Eigenvalues of the Laplacian and Invariants of Manifolds.

Sullivan, Dennis: Inside and Outside Manifolds.

Tits, Jacques**: On Buildings and their Applications.

Vitushkin, A. G.: Coding of Signals with Finite Spectrum and Sound Recording Problems.

HELSINKI (August 15-28, 1978)

Ahlfors, Lars V.***: Quasiconformal Mappings, Teichmüller Spaces and Kleinian Groups.

Calderón, Alberto P.: Commutators, Singular Integrals on Lipschitz Curves and Applications.

Connes, Alain: Von Neumann Algebras.

Edwards, Robert D.: The Topology of Manifolds and Cell-Like Maps.

Gorenstein, Daniel: The Classification of Finite Simple Groups.

Kashiwara, Masaki: Micro-Local Analysis.

Langlands, Robert P.: *L*-Functions and Automorphic Representations.

Manin, Jurii I.: Modular Forms and Number Theory.

Novikov, S. P.: Linear Operators and Integrable Hamiltonian Systems.

Penrose, Roger: The Complex Geometry of the Natural World.

Schmid, Wilfried: Representations of Semisimple Lie Groups.

Shiryaev, A. N.: Absolute Continuity and Singularity of Probability Measures in Functional Spaces.

Thurston, William P.: Geometry and Topology in Dimension Three.

Weil, André***: History of Mathematics: Why and How.

Yau, Shing-Tung: The Role of Partial Differential Equations in Differential Geometry.

WARSAW (August 16-24, 1983)

Arnold, V. I.**: Singularities of Ray Systems.

Erdös, Paul: Extremal Problems in Number Theory, Combinatorics, and Geometry.

Fleming, Wendell H.: Optimal Control of Markov Processes.

Hooley, Christopher: Some Recent Advances in Analytical Number Theory.

Hsiang, Wu-chung: Geometric Applications of Algebraic *K*-Theory.

Lax, Peter D.: Problems Solved and Unsolved Concerning Linear and Non-Linear Partial Differential Equations.

Maslov, V. P.: Non-Standard Characteristics in Asymptotical Problems.

Mazur, Barry: Modular Curves and Arithmetic.

MacPherson, Robert D.: Global Questions in the Topology of Singular Spaces.

Pelczyński, Aleksander: Structural Theory of Branch Spaces and Its Interplay with Analysis and Probability.

Ruelle, David: Turbulent Dynamical Systems.

Sato, Mikio: Monodromy Theory and Holonomic Quantum Fields – a New Link between Mathematics and Theoretical Physics.

Siu, Yum-Tong: Some Recent Developments in Complex Differential Geometry.

Plenary Lectures

BERKELEY (August 3-11, 1986)

deBranges, Louis: Underlying Concepts in the Proof of the Bieberbach Conjecture.

Donaldson, Simon Kirwan: Geometry of Four Dimensional Manifolds.

Faltings, Gerd: Recent Progress in Arithmetic Algebraic Geometry.

Gehring, Frederick W.: Quasiconformal Mappings.

Gromov, Mikhael: Soft and Hard Symplectic Geometry.

Lenstra, Hendrik W.: Efficient Algorithms in Number Theory.

Schoen, Richard M.: New Developments in the Theory of Geometric Partial Differential Equations.

Schönhage, Arnold: Equation Solving in Terms of Computational Complexity.

Shelah, Saharon: Classifying General Classes.

Skorohod, A. V.: Random Processes in Infinite Dimensional Spaces.

Smale, Stephen**: Complexity Aspects of Numerical Analysis.

Stein, Elias M.**: Problems in Harmonic Analysis Related to Oscillatory Integrals and Curvature.

Suslin, Andrei A.: Algebraic K-Theory of Fields.

Vogan, David A., Jr.: Representations of Reductive Lie Groups.

Witten, Edward: String Theory and Geometry.

ACKNOWLEDGEMENTS

We would like to express our thanks to each of the many individuals and institutions that responded to our request for photographs and information about themselves. In addition, we are in debt to each of the following for assistance of various kinds:

John and Mary Ann Addison

Geri and Lisa Albers

Godfrey Argent

Karel de Bouvere

Vladimir Drobot

Edith Fried

Terri Gutierrez

Paul R. Halmos

Penny Hill

Peter J. Hilton

Dave Jackson

Mary Jackson

Konrad Jacobs

Patricia Kenschaft

Lucien Le Cam

Edward Leigh

José A. Lino

Harry Llull

Raymond A. Macias

Saunders MacLane

Dale H. Mugler

Olga Oleinik

Ingram Olkin

Robert Osserman

Yvonne Pasos

Jean Pedersen

Stella Pólya

Marina Ratner

Dan Reid

R. M. Robinson

Betsey S. Whitman

The American Mathematical Society

Chicago Historical Society. ICHi-02228, p. 4.

The French Tourist Bureau

L'Enseignement Mathématique, 2e série, tome 1 (1955), p. 30.

Master and Fellows of St. John's College, Cambridge

Mathematical Institute of the University of Zürich

Seeley G. Mudd Manuscript Library

The Newberry Library

The New York Times (Terrence McCarthy), pp. 45, 49

Stanford University Mathematics and Computer Science Library